Fundamentals of

MODERN SCIENTIFIC COMMUNICATION

Kenneth R. Summy

Cover image © Shutterstock, Inc.

Kendall Hunt
publishing company

www.kendallhunt.com
Send all inquiries to:
4050 Westmark Drive
Dubuque, IA 52004-1840

ISBN 978-1-4652-7464-9

Printed in the United States of America

CONTENTS

Scientists engage in a variety of kinds of writing, and writing and presentations are major endeavors for scientists on a continuing basis throughout their training and careers. Unfortunately, many scientists never receive any formal training in writing and presentation. Rather, they are left to forage for scattered bits of information on their own and hope that the nuggets of wisdom they find are appropriate for the task they are working on. This text is intended as an aid to alleviate this deficiency. It provides information on how to organize various kinds of writing and presentation materials and how best to convey them to the reader or listener. It serves as an excellent aid to instruction about writing and presentation and it also serves as an outstanding reference to specific topics.

The writing is clear, lean, and precise. Examples are selected to illustrate major ideas and points in the text. They do an excellent job. The book is thorough on the kinds of writing that scientists do. It includes sections on writing Curriculum Vitae, Letters of Application (for graduate and professional schools, and jobs), Letters of Reference, Grant Proposals, and Reports of Original Research (to funding agencies and to journals). The book includes a chapter on writing review articles, books and book chapters. A chapter that one rarely sees in such books covers the importance of scientific integrity. Readers will be well informed about plagiarism and its consequences. Another unique chapter that is most welcome discusses when not to communicate information that one has created or has access to. This includes classified data, nondisclosure agreements and patent issues.

Students will find that their grades will improve as their writing and presenting skills improve, and practicing scientists will find that their careers will advance more quickly as they become adept at writing and presenting. This text will be an important tool to own whether or not one takes a formal course in writing and presenting.

Frank W. Judd, Ph.D.
Research Professor (Retired) and Professor Emeritus
The University of Texas – Rio Grande Valley

Author's Note: Dr. Frank W. Judd received his Ph.D. degree in Zoology from Texas Tech University in 1973 and served on the faculty of the Biology Department of the University of Texas – Pan American for a period of 31 years. During that time, he gained national and international recognition as both an educator and authority on the ecology of animal and plant populations in the subtropical Lower Rio Grande Valley of Texas. Dr. Judd has 133 publications including two books, and is a recipient of the Distinguished Scientist Award from the Texas Academy of Science. Following his retirement from UTPA in 2003, Dr. Judd was awarded the titles of Research Professor and Professor Emeritus, and has continued his ecological research at the newly-organized University of Texas – Rio Grande Valley in Edinburg.

Students contemplating careers in any of the natural sciences (e.g., biology, chemistry, physics, and geology) should recognize from the outset that their ultimate success in the professional world of science will be highly dependent on their ability to speak and write effectively. Application packages for various academic programs and jobs in both the public and private sectors almost invariably require applicants to submit written documents such as letters of application, *curriculum vitae* and personal statements. Qualified applicants who are invited for on-site visits to academic or research institutions must communicate effectively with interviewers, and are commonly required to present seminars of their research and teaching accomplishments to the faculty and students of the institution. Those fortunate enough to be hired for professional positions will normally be required to demonstrate productivity (in terms of presentations at scientific meetings and conferences, and publications in scientific journals) as a condition for continued employment with the institution. In short, the ability to speak and write effectively is not only a requisite for employment by a research or academic institution, but is equally important for remaining employed and advancing your career with the institution. The "publish or perish" syndrome is a reality, and students training for careers in science would be well advised to become proficient in both oral and written communication as early as possible during their educational experience.

The goal of this book is to familiarize university students who are contemplating careers in any of the natural sciences with the various types of scholarly works and other communications they will be expected to produce during their careers as practicing scientists, and to introduce some simple techniques designed to evaluate and enhance their effectiveness in public speaking and scientific writing. Part I provides an overview of modern scientific communication (Chapter 1), techniques for searching and summarizing the vast scientific literature (Chapter 2), designing, conducting and interpreting scientific surveys and experiments (Chapter 3), guidelines for reporting research effectively (Chapter 4) and concludes with a discussion of why scientific integrity is an imperative for all scientists (Chapter 5). Part II deals with the preparation of formal scientific communications including oral and poster presentations at scientific meetings and conferences (Chapter 6), journal articles and other reports of original research (Chapter 7), books, book chapters, and review articles (Chapter 8) and concludes with a discussion of the publication process for scientific manuscripts and other documents (Chapter 9). Part III covers three important documents—Letters of Application, Curriculum Vitae and Personal Statements—that are normally included in most job application packages (Chapter 10), proposals for grants and contracts (Chapter 11), letters of recommendation for students and colleagues (Chapter 12). Part IV covers some tricky issues associated with intellectual property, copyrighted materials and classified materials (Chapter 13) and concludes with a discussion of why it is advisable for students to "remain students for the remainder of their lives" with some advice on how to maintain your credibility and stance in the scientific community once you have achieved your career goals (Chapter 14).

ACKNOWLEDGMENTS

I would like to thank Dr. Robert Edwards and Dr. Frank Judd for critical reviews of this book, and all of the faculty in the UTRGV Biology Department who have provided support for this project and who have worked so diligently to establish an effective scientific communications program at this university. Finally, I would especially like to thank my wife Laura for her patience, support and reviews while the writing was in progress.

Basic Concepts and Skills

Modern Scientific Communication

The Dissemination of Scientific Knowledge

Chapter Learning Objectives

After studying this chapter, you should understand:

- The fundamental difference between *science* and *technology* and how the two are linked by *scientific communications*.
- The various means by which new scientific knowledge is disseminated to the scientific community and to the rest of the world.
- Why it is imperative that students contemplating careers in the natural sciences become proficient in both oral and written communications as early as possible during their academic training.

Science and Technology in Modern Society

Although the terms *science* and *technology* are often used interchangeably, it is important to recognize an important distinction between the two. Science is derived from the Latin words *scientia* ("knowledge") and *scire* ("to know)" and refers to a systematic and rational approach to investigating natural phenomena through a combination of observations, experimentation, and testing of tentative explanations or *hypotheses,* all of which are elements of the *scientific method* (Gower, 2002).[1] Science is a philosophy that seeks to gain new knowledge of natural phenomena regardless of whether or not any practical applications for such knowledge currently exist (to some people, this almost seems like seeking new knowledge for the sake of gaining new knowledge). In contrast, *technology* refers to the development of practical applications and/or products from new scientific knowledge that will hopefully be profitable for the developer(s) and beneficial to society as a whole (unfortunately, this is not always the case). A brief historical review of three modern industries should exemplify how the interaction between science and technology has resulted in products and developments that have fundamentally changed human societies (Evans, 1959; Poe, 2011; Young, 1998).

Telecommunications. Early basic research by Henry, Faraday, and others on the nature and behavior of electricity and electromagnetism (examples of pure science) led to the invention of the telegraph in 1831, the telephone in 1870, the radio in 1905, and television in 1925 (examples of technology in action)[2] (Rooney, 2013). By the early 20th century, communications conglomerates such as American Telephone and Telegraph (AT&T) had formed and established large communications networks in many areas of the developed world. Later research on semiconductors (metals with properties intermediate between conductors and insulators) conducted by Bell Laboratories during 1946–1947 led to the development of the transistor,[3] which provided the basis for mobile telephone services which are now used by millions of people worldwide, and triggered a revolution in electronics and computer technology.

Computer Technology. Early research on computing led to the development of a series of mechanical computers between 1834 and 1890 and the Tabulating Machine Company, which would later become International Business Machines (IBM) was founded during 1896[4]. A variety of computer languages were developed during 1931–1937, and a calculator that would provide the basis for digital computers was developed during 1939. Several major computer systems were developed and used by both sides during World War II for either encryption (*Enigma* by Germany) or code-breaking and other activities (the *Collosus* series by Great Britain; the *Mark* series and ENIAC by the United States). During the postwar period, research on computer technology accelerated and led to the development of several high-speed computer systems (UNIVAC, EDVAC, IBM701), new computer languages (FORTRAN, COBOL, and BASIC), and the world's first supercomputer (CDC 7600) by 1969.

A continuation of research and development of computer technology led to the development of integrated circuits in 1950 and microprocessors in 1970 which provided the basis for the production of desktop and laptop computers that are now used by millions worldwide. In turn, the enormous demand for personal computers generated a major software industry which has developed

[1] Wikipedia, "History of the Scientific Method," *Online*, http://en.wikipedia.org/wiki/History of scientific method (April 2, 2015).

[2] ShoreTel, "The History of Telecommunication," *Online*, https://www.shoretel.com/history-telecommunication (April 4, 2015).

[3] ScienCentral (PBS), "Transistorized," *Online,* http://www.pbs.org/transistor (April 5, 2015).

powerful word-processing software (e.g., MicroSoft® Word), presentation software (e.g., MicroSoft® Power Point), user-friendly statistical packages (e.g., SPSS,® SAS,® and Minitab®) and other products that have greatly simplified the analysis of data and preparation of oral and written scientific communications.[4]

The Internet and World Wide Web. One of the most significant achievements in the history of science and technology was the development and implementation of the *Internet* and *World Wide Web*.[5] A continuation of research on electronic computer technology and telecommunications that had begun during the 1950s resulted in the development of "*packet-switching*" networks (e.g., ARPANET) capable of transmitting data over long distances by 1969 (packet switching technology breaks transmitted messages into separate packets which are transmitted separately by several means to the destination). Further research led to the development and adoption of standard protocols for data transmission that allowed separate networks to be merged into "networks of networks" (e.g., TCP/IP became the standard protocol for ARPANET by 1982). The establishment of NSF-funded supercomputing centers at several universities in the United States resulted in the establishment of additional interconnected networks (CSNET, NSFNET), which were available primarily to the military, researchers, and educators. With the development of the World Wide Web during the 1980s and the elimination of restrictions on commercial uses during the 1990s, a global telecommunications network was established and the Internet (also known as the "Information Superhighway") was born and continues to grow in importance.

Scientific Communication—The Link Between Science and Technology

If the research that led to the development of modern computer and telecommunications systems had been forever classified as military secrets and communication among the scientists and technologists who were involved in the various research projects had been stifled for whatever reason, the computer and telecommunications revolution that we are experiencing now would never have occurred (or at least would have been delayed for many years). Similarly, if communication and interaction among medical scientists is ever eliminated or severely restricted by political or other means, efforts to prevent and/or contain major outbreaks of human diseases will be curtailed substantially, and epidemics such as those which have ravaged widespread areas in many areas of the world during the past several decades may be expected to continue indefinitely. Scientists continue to make remarkable discoveries in medicine and other disciplines on a routine basis, and the development of practical applications and useful products from this knowledge is predicated on the ability and willingness of scientists to communicate new discoveries to other scientists and to the rest of the world by various means.

Conventional Communications. Like most other members of society, scientists routinely communicate with each other via telephones, mail services, and discussions during *ad hoc* meetings or teleconferences. These conventional communications are particularly important during the planning phases of research projects and in many cases, during ongoing projects. It is important to recognize

[4] Siteseen, Ltd. "Computer History Timeline," *Online,* http://www.datesandevents.org/eventsptimelines/07/-computer-history-timeline.htm (March 4, 2015).

[5] Wikipedia, "History of the Internet," http://en.wikipedia.org/wiki/History of the Internet (May 4, 2015).

that significant changes are occurring in several of these technologies due to rapid developments that continue to occur in the computer and telecommunications industries:

- **Cell Phones** with more functional capabilities than most early desktop computers have become increasingly competitive with conventional landline telephone systems, and facilitate wireless communications over very long distances provided that appropriate relay facilities are available and operational.

- **E-Mail** has become highly competitive with conventional mail services ("snail mail") and provides almost instant access to e-mail addresses worldwide. Moreover, e-mail messages may be sent and received via cell phones, laptops, and other computer devices with Internet connections.

- **Teleconferencing** technology has become widely available and provides a viable alternative to conventional meetings that must be attended personally and typically involve travel to distant locations (one exception to this involves annual meetings of scientific societies which are rarely conducted via teleconferencing). Use of teleconferencing software such as Skype® (www.skype.com), Web-X® (www.webx.com), AnyMeeting® (www.anymeeting.com), and many others allows ad hoc meetings to be conducted on short notice and without the travel expenses that are typically associated with conventional meetings.

The Scientific Media. Once a study or experiment has been conducted, analyzed and interpreted, the final critical step in completing a research project is to disseminate the new scientific information to other scientists and to the general public via one or more of the following means:

- **Press Releases**—preliminary reports of research that may be broadcast to the public via radio or television, or published in newspapers, magazines, newsletters, websites, and similar media.

- **Oral Presentations and Posters**—these are the two principal means for disseminating new information at scientific meetings and conferences, and are typically presented during timed sessions. Oral and poster presentations may involve either preliminary results of current research (in progress) or recently completed studies. In many instances, abstracts of oral and poster presentations are published in a Conference Proceedings for distribution to conference attendees.

- **Theses and Dissertations**—these are summaries of research conducted by graduate students at academic institutions to fulfill requirements of Master of Science (M.S.) or Master of Arts (M.A.) degrees and Doctor of Philosophy (Ph.D.) degrees. In most cases, research reported in theses and dissertations will be published as research articles in scientific journals.

- **Unpublished Research Reports**—this category includes *laboratory reports* for undergraduate and graduate courses at academic institutions, and internal documents used by government agencies and private institutions to document research accomplishments involving highly sensitive or classified data which cannot be released to the public at the present time.

- **Research Articles in Scientific Journals**—journals are the principal means by which reports of original research (i.e., the *primary scientific literature*) are disseminated to the scientific community. One major distinction between journal articles and other scientific media is that the former are normally subjected to rigorous peer reviews by anonymous experts selected by journal editors prior to publication (the manuscript peer review process will be discussed extensively in Chapter 9).

- **Books, Book Chapters, and Review Articles**—these publications typically consolidate and summarize results of previous research on selected topics and hence, are known as the *secondary* scientific literature. Although review articles published in journals are typically subjected to anonymous peer reviews, the review process for book chapters and books may vary considerably among publishers.

A Tale of Three Students—What You Can Expect in the Real World

The following career profiles of three hypothetical students should also illustrate why learning to speak and write effectively is not only desirable, but is essential for anyone contemplating a career in the natural and physical sciences. Following their graduation from high school, John Q. Public, John Doe and his sister Jane took their SAT tests and submitted applications for admission to Hypothetical University (HU) as Biology majors and Chemistry minors. During the next four years, all three students enrolled in several writing and seminar courses and gained additional experience in scientific communication by delivering oral presentations and writing laboratory reports in several of their courses.

Following their graduation from HU with Bachelor of Science degree, John Doe applied for a technical position that had been advertised by a major university (Figure 1.1). Requirements for the position included a B.S. degree in one of the natural sciences with a preference for candidates with previous training in basic statistics and scientific communications. The salary was lucrative and the working conditions were great, so John accepted the position and began a professional career that would involve a considerable amount of public speaking and technical report writing.

Position Announcement
Research Technician

The Department of Environmental Sciences at the University of Hypothetical Studies seeks to fill a Research Technician position beginning as soon as position is filled (Job Vacancy No. FY13/016).

Minimum education requirements: B.S. degree in Environmental Science or related discipline, with preference for candidates with previous experience in use of statistical software and proficiency in scientific communications. Information on the UHS Environmental Science Department can be obtained at www.UHS.EnvSci.edu. The selected candidate will assist faculty in the development of field and laboratory experiments, data analysis, and writing internal reports.

Figure 1.1 Excerpt from a position announcement posted by a major academic institution indicating the qualifications for the position and responsibilities of the applicant hired.

John Q. Public and Jane Doe wanted to continue their education, so both took their GRE tests and submitted their application packages (which included a letter of application, *curriculum vitae* (CV), and personal statement) to the graduate school at HU and both were admitted into the Environmental Science graduate program. During their first semester as graduate students, John and Jane enrolled in several environmental science courses, all of which required oral presentations and/or written lab reports, and both conducted exhaustive literature searches of their chosen thesis research topics (John's topic was "ecology of parasites associated with wood rats" and Jane's was "effects of photoperiod on food consumption by German cockroaches) using on-line databases available in the university library. During the next two years, John and Jane conducted well-designed studies and made several oral presentations at scientific meetings and conferences, wrote manuscripts that were submitted to a prestigious refereed journal and each completed a *thesis* of their research projects that was accepted by their graduate advisory committees and the university graduate school. During their graduate studies, both had also gained teaching experience by serving as laboratory teaching assistants (TAs) in several undergraduate courses, and were awarded scholarships by a scientific society to support their research projects. All of these accomplishments were entered into their rapidly expanding curriculum vitae (CVs).

John and Jane graduated from HU with Master of Science (M.S.) degrees, and Jane applied for a Lecturer position at a prestigious university (Figure 1.2). The position required an M.S. degree with demonstrated proficiency in both teaching and scientific communications. The salary was handsome and the working conditions were great, so she accepted the position and began a career that would involve a combination of science and journalism for the next several decades.

John Q. Public decided to continue his education, so he submitted an application for admission to the doctoral program at Theoretical University (TU), one of the most prestigious universities in the country, and was admitted to their Ph.D. program.

During the next two years, John completed additional coursework, taught laboratory sections for several courses, published several papers based on his M.S. research, and conducted an extensive study of wood rat ecology that was summarized in a *dissertation* that was approved by his graduate advisory committee and the university administration. John graduated with a Doctor of Philosophy (Ph.D.) degree in environmental science and shortly thereafter began a job hunt in earnest.

Position Announcement
Lecturer

The Department of Geology at the University of Theoretical Research seeks to fill a Lecturer position beginning as soon as position is filled (Job Vacancy No. FY15/015).

Minimum education requirements: M.S. degree in Geology or a related discipline, and a demonstrated record of accomplishments in both teaching and scientific communications. Information on the Geology Department at UTR can be obtained at www.UTR.Geology.edu. The selected candidate will develop and teach courses in Geology and Scientific Communications.

Figure 1.2 Excerpt from a position announcement (Lecturer) posted by a major academic institution indicating the qualifications for the position and responsibilities of the applicant hired.

Position Announcement
Assistant Professor—Ecology

The Department of Environmental Science at Theoretical University seeks to hire an Assistant Professor specializing in ecology beginning as soon as position is filled (Job Vacancy No. FY17/017). Minimum requirements: Ph.D. degree in Ecology with a demonstrated record of teaching and research. Information on the Biology Department at TU can be obtained at www.TU.Biology.edu. The selected candidate will be responsible for developing graduate and undergraduate courses in Ecology and develop a productive research program that addresses significant problems and employs students in research.

Figure 1.3 Excerpt from a position announcement posted in a public medium by a major institution indicating the qualifications and responsibilities of the position.

Dr. John Q. Public submitted a number of job applications to research and academic institutions, each of which contained statements of teaching and research interests and a copy of his now-impressive and burgeoning CV. He was interviewed by phone by two institutions and was invited for an on-site interview by one of these (Figure 1.3). During the interview, he met most of the faculty and administrators and delivered a 30-minute seminar of his research accomplishments and future research plans to the faculty and students of the institution. Shortly thereafter, he was offered the position at a lucrative salary and a large start-up fund to establish his research program. Dr. Public signed a contract with TU and was appointed as a tenure-track Assistant Professor in the Environmental Science Department. He later learned that the competition for the position he had been selected for had been intense—35 Ph.D.s from across the country had originally applied for the position, three had been invited for on-site interviews and one (Dr. Public) was selected for the position.

After celebrating his victory, Dr. Public wasted no time in developing his teaching and research programs. As a *tenure-track* Assistant Professor, he was responsible for teaching lecture and laboratory sections for each of two courses per semester, and for developing a vigorous research program in environmental science (Figure 1.4). He was aware that the requirements for *tenure* (award of permanent job status) at this particular institution mandated that he publish a minimum of three scientific articles in refereed journals and deliver a minimum of five oral presentations and/or posters at regional, state, national or international conferences during his 5-year probationary period. In addition, he was expected to demonstrate a high degree of effectiveness in teaching activities and to submit proposals for grants and contracts from various funding sources in an effort to fund his research program, which would eventually include undergraduate and graduate student researchers. Dr. Public succeeded in fulfilling these requirements and was awarded tenure and promotion to the rank of *Associate Professor* at the end of his 5-year probationary period. He had "proved" himself to his colleagues and had finally attained permanent job status on the HU faculty. Had he not done so, his position would have been terminated at the end of the academic year and he would have been back where he was five years earlier.

The story does not end here. Most if not all reputable research and academic institutions expect tenured faculty (Associate and Full Professors) to continue their scholarly productivity as long as they are employed by the institution(s). Dr. Public succeeds in meeting these requirements and is

Department of Biology
Theoretical University
Tenure and Promotion Requirements

Award of Tenure and Promotion to Associate Professor. Applicant must have a minimum of five years teaching experience as Assistant Professor and a minimum of three (3) publications in refereed journals and five (5) presentations at scientific meetings and conferences. Applicant must have acceptable ratings on student evaluations in all courses taught during probationary period.

Promotion to Professor. Applicant must have a minimum of ten (10) years total teaching experience and a minimum of ten (10) publications and ten (10) presentations at scientific meetings and conferences during his or her entire academic career. Applicant must also have acceptable ratings on student evaluations on a minimum of 90% of courses taught during the evaluation period.

Figure 1.4 Tenure and Promotion Requirements for a theoretical Department in a Theoretical University.

promoted to the rank of Full Professor five years later. By this time, Dr. Public has become a well-known expert in his field and has numerous undergraduate and graduate students working under his supervision. This is a particularly rewarding aspect of his job as he realizes he is training and mentoring the future scientists who will eventually replace him and his contemporaries in the scientific world. At this point in his career, he has also assumed many other responsibilities that involve speaking and writing—he is routinely contacted by editors of scientific journals to review submitted manuscripts, and by managers of various funding sources to review proposals for grants and contracts submitted by scientists throughout the country. In addition, many of his former students routinely contact him with requests for letters of reference that will help them get their own careers off the ground.

One day, Dr. Public announces his intent to retire and submits the required paperwork to the university administration. His retirement party is attended by colleagues from his own institution and others around the country, including both John and Jane Doe and a considerable number of his former students who are now accomplished professional scientists and educators themselves. They present him with a prestigious award and after his acceptance speech, give him a 1-minute standing ovation. When asked what he intended to do during his retirement, Dr. Public replied that he planned to continue writing books and book chapters, review articles and other types of scientific communications relating to his discipline. In addition, he planned to accept many of the requests for invitational seminars and symposia that he had begun receiving since his official retirement announcement. The only sour note of the day occurred when a former student informed him that one of his colleagues from graduate school had lost his job and ruined his career because of a *plagiarism* charge (a detailed discussion of plagiarism and other forms of academic misconduct are provided in Chapter 5). The bottom line is that Dr. Public enjoyed a long and highly productive career as a scientist largely because he had learned how to speak and write effectively early in his career, and had maintained a high level of productivity and an impeccable level of professional integrity throughout his 30-year career. Somewhere along the line, he had also learned to enjoy speaking and writing.

Some Sage Advice for Prospective Scientists—How to Survive the "Publish or Perish" Syndrome

The previous discussion of John Q. Public and his colleagues is fictional but very typical of the career profiles of many professional scientists and educators. Prospective scientists must demonstrate the ability to speak and write effectively in order to get jobs with research and academic institutions, and must routinely deliver oral presentations and publish articles in journals and other media in order to keep those jobs and advance in their careers (this is the familiar "*publish or perish*" syndrome you have probably heard about). In addition to these activities, professional scientists (like many professionals in other fields) are generally responsible for handling a daily flow of memoranda, reports and other important documents, many of which have very short deadlines. As long as the scientist is competent in scientific communication skills and enjoys his or her work, these chores tend to be rewarding and the phrase "publish or perish" is usually regarded as sage advice rather than a veiled threat.

Your take-home message is this—if you are planning to launch a career in one of the natural sciences (or many other fields of study), you must become proficient in both oral and written communication skills if you expect to become successful, and the sooner this is accomplished the better. If after completing this course, you dislike writing intensely and have a morbid fear of public speaking, you might want to reconsider your career plans. Otherwise, study the materials contained in the remaining chapters of this book (or other similar books) and become thoroughly familiar with the various types of scientific presentations, publications, and other documents that natural scientists must deal with routinely. Once you become proficient and productive in scientific communications, you can reasonably expect to have a successful career and to enjoy your work as time progresses.

Exercise 1.1

Visit the website of one or more major universities and navigate to one of the departments in natural sciences (biology, geology, physics or chemistry). Click on the faculty tab and review the CVs of several faculty members, with particular emphasis on the types of courses taught and the professional accomplishments listed (presentations, publications, grants, and other accomplishments). Then go to the section(s) on positions available and compare (1) the qualifications required for positions with various levels of complexity (technicians, lecturers, and tenure-track positions in various disciplines) and (2) the duties relating to speaking and writing for each of these positions. This exercise should clarify what will be expected of you if you apply for a research or academic position, and what your expectations for speaking and writing will be if you accept such a position.

References

Rooney, A. (2013). *The story of physics*. London: Arcturus, 208 pp.

Evans, H. M. (1959). *Men and moments in the history of science*. Seattle: University of Washington Press, 226 pp.

Gower, B. (2002). *Scientific method: A historical and philosophical introduction*. New York: Taylor & Francis e-Library, Routledge, 276 pp.

Poe, M. T. (2011). *A history of communications: Media and society from the evolution of speech to the internet*. New York: Cambridge University Press, 337 pp.

Young, G. (1998). *The internet*. New York: H. W. Wilson, 215 pp.

Test Your Knowledge

Without referring to the material presented earlier in this chapter, mark an "x" in the box of the most appropriate answer to each question. Answers are provided at end of chapter.

1. The term *science* is derived from the Latin term for ___________________.
   - ❑ **a.** "study"
   - ❑ **b.** "knowledge" or "to know"
   - ❑ **c.** "experiment"
   - ❑ **d.** "observation"

2. When and where did the *scientific method* first become widely adopted by the majority of practicing scientists?
   - ❑ **a.** ancient Greece ~ 2,000 B.C.
   - ❑ **b.** Rome ~ 350 A.D.
   - ❑ **c.** 17th century Renaissance in Europe
   - ❑ **d.** 18th century Industrial Revolution in Europe and North America

3. The *scientific method* includes all of the following EXCEPT
   - ❑ **a.** simulation by modeling
   - ❑ **b.** experimentation
   - ❑ **c.** systematic observation and data collection
   - ❑ **d.** hypothesis testing

4. *True or False*. Because of the high costs of research, most modern experiments are conducted to develop new products or improve existing ones.
   - ❑ **a.** True
   - ❑ **b.** False

5. All of the following are examples of *pure science* EXCEPT
   - ❑ **a.** Kepler's laws of planetary motion
   - ❑ **b.** Newton's law of gravitation
   - ❑ **c.** Henry and Faraday's research on electromagnetic properties of metals
   - ❑ **d.** Lippershey's development of the telescope

6. What is the primary role of *technology* in modern society?
   - ❑ **a.** the design of complex instrumentation
   - ❑ **b.** to provide the basis for scientific experimentation
   - ❑ **c.** to explain complex natural phenomena
   - ❑ **d.** to utilize scientific knowledge for development of useful products

7. All of the following represent classic examples of technology in action EXCEPT

❑ **a.** invention of telephone

❑ **b.** development of wireless communications

❑ **c.** research on "black holes" in space

❑ **d.** development of the Internet and World Wide Web

8. FORTRAN, COBOL and BASIC are examples of ____________________.

❑ **a.** wireless communications technology

❑ **b.** computer programming languages

❑ **c.** computer systems used in the World Wide Web

❑ **d.** encryption technologies for wireless transfer of data

9. Which of the following is the primary means by which new scientific findings are disseminated to the scientific community and general public?

❑ **a.** unpublished reports

❑ **b.** review articles

❑ **c.** press releases

❑ **d.** refereed journals

10. *True or False.* Despite the numerous changes that have occurred in the communications industry during the past three decades, students contemplating careers in the natural sciences (and numerous other fields) must develop proficiency in speaking and writing if they expect to be successful in the professional world of science.

❑ **a.** True

❑ **b.** False

Answers: 1) b; 2) c; 3) a; 4) b; 5) d; 6) d; 7) c; 8) b; 9) d; 10) a.

Using the Scientific Literature

Techniques for Locating, Summarizing, and Citing Scientific Articles

Chapter Learning Objectives

After studying this chapter, you should understand:

- How to conduct efficient literature searches using library catalogs, online databases, search engines and other resources available in most university libraries.
- How to summarize published scientific information and conduct meaningful comparisons of studies conducted by various researchers.
- How to properly cite literature sources in oral and written communications.

One of the first tasks in developing new research programs is for the researcher(s) to become thoroughly familiar with what is currently known about the research topic that will be addressed. By comparing similarities and differences of research conducted previously by different researchers, important gaps in knowledge may be identified which may provide the basis for one or more objectives of the proposed research program. In order to accomplish this, it is necessary for the researcher(s) to have available most or all of the scientific literature dealing with the topic of interest.

Literature Searches

The purpose of a literature search is to locate and assemble scientific literature that is important and relevant to a research problem under investigation. Since the scientific literature is vast and consists of untold numbers of articles in thousands of journals, books and other sources located in all areas of the world, it is imperative that literature searches be conducted in a systematic manner that is both efficient and effective (in particular, the search should detect all major works relating to the problem in question). This requirement provides a strong incentive for researchers to become thoroughly familiar with the resources available in their libraries and on various websites of the Internet.

Library Resources. Prior to the advent of the computer age, literature searches were most commonly conducted by (1) obtaining reprints (paper copies) or photocopies of relevant articles from recent issues of journals and other primary literature, (2) poring through paper volumes of abstracting journals such as *Biological Abstracts* and *Chemical Abstracts Service* (CAS) for articles relevant to the proposed research topic, (3) locating books and other secondary literature available in their libraries using traditional (paper) card catalogues, and obtaining references that were not available locally through interlibrary loan services, and (4) using *Science Citation Index* to locate more recent articles that cited older papers on the topic in question. Additional literature on the research topic was commonly located by examining the *Literature Cited* sections of available publications for articles that were not detected using the other four techniques. These methods were used by scientists for many years and were generally effective, but tended to be time-consuming and laborious.

Today, the traditional (paper) card catalog system has been computerized in most libraries and provides almost instant access to most or all of the library's resources, including journals, books, government documents, special collections and archives, and numerous other sources of information. By simply entering one or more key words (e.g., author, subject, title), the user is provided with a list of relevant materials that are available in the library, or which may be obtained from other facilities through the interlibrary loans services (Figure 2.1).

Online Databases. During the past several decades, much of the recent scientific literature has been digitized and is now archived in on-line databases. A visit to the website of any major university library will reveal hundreds of on-line databases each containing publications from selected journals or other sources (Figure 2.2). Many of these databases cover a wide range of disciplines in the natural sciences, e.g., according to the menu mentioned previously, *Science Direct* ". . . provides full image articles from more than 1000 agriculture, arts and humanities, astronomy, biology, business, chemistry, clinical medicine, computer science, earth and planetary sciences, economics, engineering, energy and technology, environmental science, life sciences, materials science, mathematics, physics, and social and behavioral sciences journals." *Web of Science* ". . . contains three citation databases—*Science Citation Index Expanded*, *Social Sciences Citation Index*, and *Arts & Humanities Citation Index*, which include citations and some links to full text articles from 1975 to the present."

Catalog | Library U Search | Journal Title | Subject Guides

For ***Advanced Keyword*** search, click Here.

Subject Heading | geomorphology | Search

Español

More...
ILL Request Form
Reserves
Reserve Request Form

Save Marked Records | Save All On Page

Num	Save	SUBJECTS (1-50 of 53)	Year	Entries 111 Found
1		Geomorphology -- 8 Related Subjects		8
2	☐	Geomorphology		42
3	☐	Geomorphology Arid Regions : Mabbutt, J. A.	1977	1
4	☐	Geomorphology Arkansas Buffalo River Watershed : Panfil, Maria S.	2001	1
5	☐	Geomorphology Asia : Kuhle, Matthias.	c2013	1
6	☐	Geomorphology Assateague Island Md And Va Maps : Morton, Robert A.	2008	1
7	☐	Geomorphology Australia : Holdaway, Simon J.,	c2014	1
8	☐	Geomorphology Australia Northern Territory : Wohl, Ellen E.,	c1994	1
9	☐	Geomorphology Australia Queensland : Wohl, Ellen E.,	c1994	1
10	☐	Geomorphology California Manix Lake Guidebooks : Reheis, Marith C.	2007	1

Figure 2.1 Computerized card catalog of a modern university library (upper) provides almost instant access to the library's resources (lower).

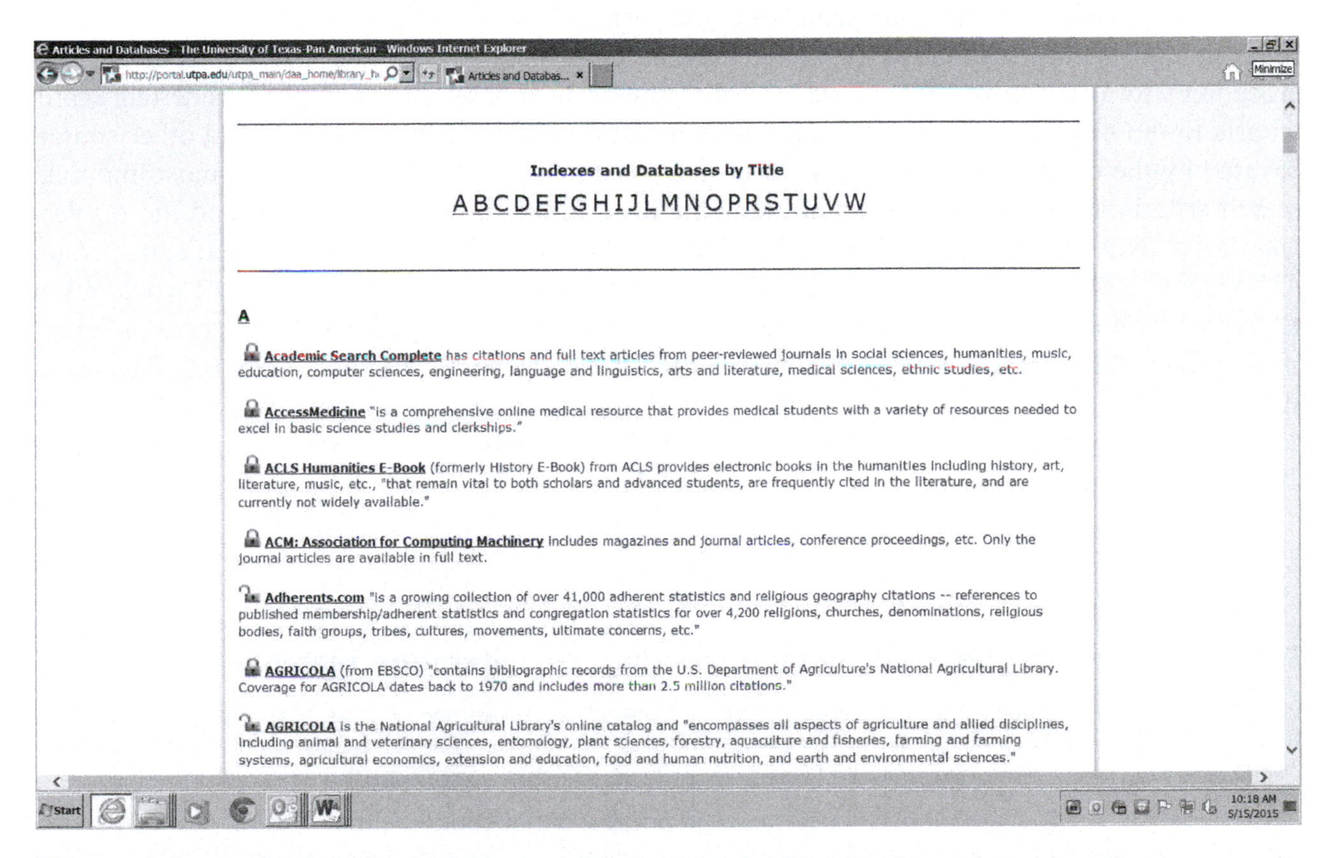

Figure 2.2 Menu of on-line databases maintained by the library of a major university.

Many other databases are more restricted in scope (e.g., *PubMed* is described as "... a service of the National Library of Medicine, provides access to over 11 million MEDLINE citations back to the mid-1960's and additional life science journals. PubMed includes links to many sites providing full text articles and other related resources)." *Biological Abstracts 1969-Present* "... covers international journals in the life sciences." AGRICOLA (from EBSCO) "contains bibliographic records from the U.S. Department of Agriculture's National Agricultural Library. Coverage for AGRICOLA dates back to 1970 and includes more than 2.5 million citations." *ChemWeb* is "the largest online chemical community in the world... combines a huge range of information for those in research chemistry, the chemicals industries and related disciplines." *Naturephysics* publishes [full text] papers of the highest quality and significance in all areas of physics, pure and applied ... [including] core physics disciplines, but also a broad range of topics whose central theme falls within the bounds of physics."

Several on-line databases are highly specialized and provide information on literature sources that are not normally listed in databases specializing in refereed journals. *Dissertations and Theses* (Dissertation Abstracts) is "... the single, authoritative source for information about doctoral dissertations and master's theses... from over 1,000 graduate schools and universities." *Books in Print* "... provides title, author, ISBN number, and publisher information for millions of books." *GPO Access* "... is a service of the U.S. Government Printing Office that provides the official, published version of important information produced by all branches and agencies the Federal Government." Other databases provide specialized services such as journal rating indexes (e.g., *Journal Citation Reports (JCR): Science Edition* is "... the leading authority for evaluating the impact of cited references in science and indexes over 5,900 international science journals)." Collectively, these on-line databases are user-friendly and provide access to virtually all of the published literature required for conducting and reporting research in the natural sciences, and have largely eliminated the drudgery that was associated with literature searches in the past.

Although the information sources covered by the various online databases varies considerably, the procedures for using most of them are basically similar. In general, the user specifies certain search criteria to the database which in turn, provides a report of articles from journals and other sources covered by the database which satisfy the criteria specified by the user. Two of the most important search criteria are *keywords* (words or phrases related to the topic of the search) and the *Boolean operators* AND, OR, and NOT (Table 2.1). When two keywords are separated by a Boolean operator, AND would report all articles within the database containing both of the keywords, OR would report all that contain either or both keywords (and hence, would be much larger than the previous report) and NOT would report all articles that include one of the keywords but not the other(s). The use of appropriate keywords in conjunction with Boolean operators provides an effective means to narrow down a literature search to a manageable number of relevant publications in a much shorter time period than was required in the past.

Table 2.1 Boolean operators and their use in online literature searches.

Operator	Function
AND	Reports all literature in database that contain both keywords
OR	Reports all literature in database that contain either or both keywords
NOT	Reports all literature that contains one keyword, but excludes the second

Example 2.1—Using Online Databases

A medical scientist interested in developing a new research program on dengue fever (a tropical disease vectored by mosquitoes) initiates a literature search by entering the keyword "dengue" in a subject box on the home page of the online database Science Direct (Table 2.2). When the search button is clicked, the user is provided with a list of 18,335 references, each of which relates to the topic "dengue" without regard to the specific subject matter or geographic location on earth. This is far too many references, so the researcher narrows the topics somewhat by entering the criteria "dengue fever AND mosquitoes" which reduces the list to 7,553 references which relate to both dengue fever and mosquito vectors regardless of geographic location. The researcher narrows the topics further by entering "dengue AND mosquitoes AND Texas" which further reduces the list to 927 articles that deal with dengue fever and their mosquito vectors in the state of Texas.

Finally, the keywords "dengue AND mosquitoes AND Texas AND Lower Rio Grande Valley" provide a list of 27 articles that deal with dengue and its mosquito vectors in a restricted four-county region of extreme southern Texas. Each of these articles is provided with a downloadable abstract which can be viewed (and copied) to determine its relevance to the proposed research, and many contain a PDF file of the full text which can be downloaded with or without charge (Figure 2.3).

Search Engines. In addition to online databases, scientific literature relating to a proposed research topic may be located by using one or more of the web search engines [e.g., Google (www.google.com), Bing (www.bing.com), Yahoo!Search (www.search.yahoo.com), Ask (www.ask.com), AolSearch (www.aolsearch.com)]. In contrast to the online databases for literature, web search engines provide access to a much wider range of resources (e.g., literature, imagery, videos) from a much wider range of sources (e.g., blogs and websites of public agencies, private organizations and businesses). Thus, a Google search based on the keywords "dengue AND mosquitoes AND Texas AND Lower Rio Grande Valley" reported 7,280 references in 0.88 seconds, which was far greater than the 27 references reported by Science Direct using identical keywords and operators (Figure 2.3). This discrepancy was due to the fact that the Google® report included not only refereed journal articles and other scientific literature, but also contained an abundance of imagery, videos, reports from public and private websites, and other sources. The reader should note that Google Scholar® is also available and is designed to search databases of scientific literature similar to those used by Science Direct and other databases discussed previously and hence, reports from this search engine will normally be considerably smaller than those based on a conventional Google® search.

Collectively, the combined use of online databases and web search engines provides the means to rapidly locate and assemble most or all of the literature and other materials required for scientific communications from virtually all locations on earth accessible by the Internet.

Table 2.2. Reports generated by selected keywords separated by Boolean operators.

Keywords and Operators	Citations
dengue	18,335
dengue AND mosquitoes	7,553
dengue AND mosquitoes AND Texas	927
dengue AND mosquitoes AND Texas AND Lower Rio Grande Valley	27

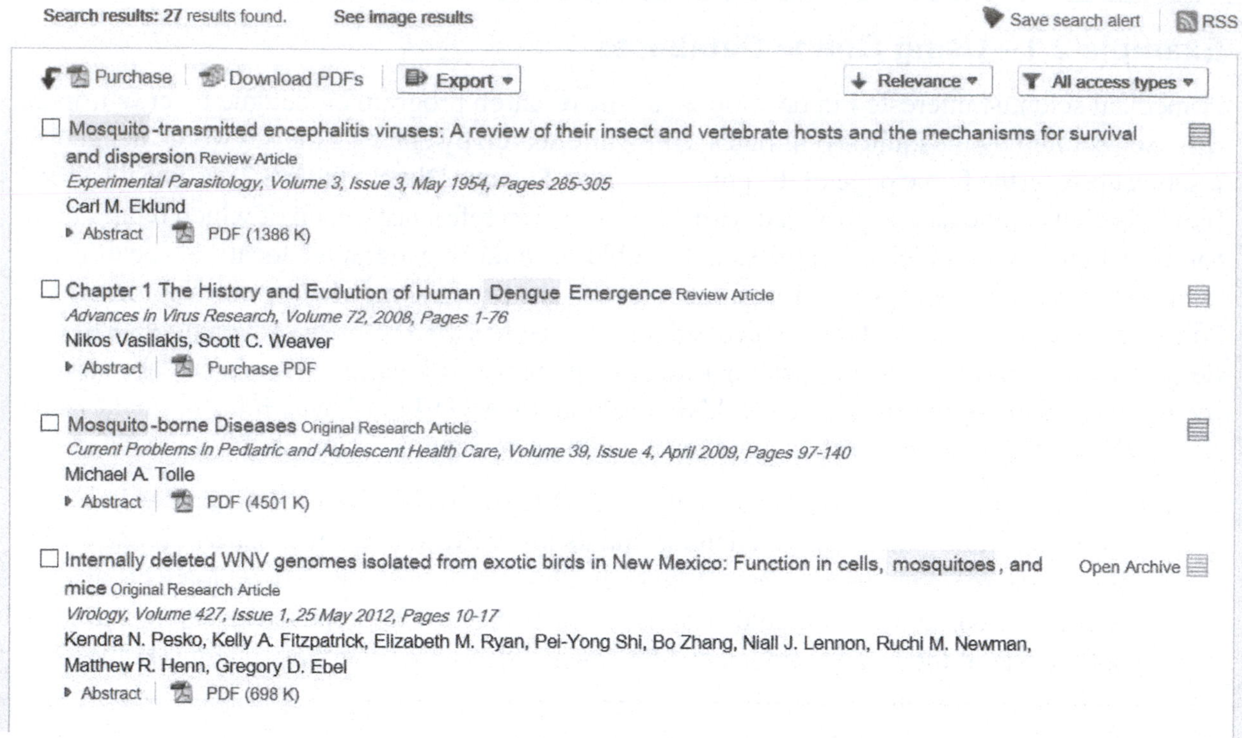

Search results: 27 results found. See image results Save search alert RSS

Purchase Download PDFs Export Relevance All access types

Mosquito-transmitted encephalitis viruses: A review of their insect and vertebrate hosts and the mechanisms for survival and dispersion Review Article
Experimental Parasitology, Volume 3, Issue 3, May 1954, Pages 285-305
Carl M. Eklund
Abstract | PDF (1386 K)

Chapter 1 The History and Evolution of Human Dengue Emergence Review Article
Advances in Virus Research, Volume 72, 2008, Pages 1-76
Nikos Vasilakis, Scott C. Weaver
Abstract | Purchase PDF

Mosquito-borne Diseases Original Research Article
Current Problems in Pediatric and Adolescent Health Care, Volume 39, Issue 4, April 2009, Pages 97-140
Michael A. Tolle
Abstract | PDF (4501 K)

Internally deleted WNV genomes isolated from exotic birds in New Mexico: Function in cells, mosquitoes, and mice Original Research Article Open Archive
Virology, Volume 427, Issue 1, 25 May 2012, Pages 10-17
Kendra N. Pesko, Kelly A. Fitzpatrick, Elizabeth M. Ryan, Pei-Yong Shi, Bo Zhang, Niall J. Lennon, Ruchi M. Newman, Matthew R. Henn, Gregory D. Ebel
Abstract | PDF (698 K)

Figure 2.3 Partial list of publications relating to vectors of dengue fever in south Texas reported in a literature search using an online database. Notice provisions for downloading abstract and/or the publication itself.

Social Networks for Scientists

Routine literature searches using online databases provide one means for scientists and educators to keep abreast of the rapidly changing literature of their respective disciplines. An effective complement to online databases are several *social networks for scientists* such as ResearchGate® (www.researchgate.net), LinkedIn® (www.linkedin.com), Academia.edu® (www.acasdemia.edu) and others. Social networks are used extensively by scientists in many disciplines, and provide a forum for the sharing of ideas and exchange of information, in addition to providing the opportunity for scientists to greatly expand their professional contacts and relationships. Most social networks maintain online databases of your publications which are updated routinely as new publications become available. In addition, most of these networks contain information regarding the ratings of journals you publish in and citations of your work by other scientists. Both of these factors are extremely useful in the preparation of job application packages, personnel evaluations and applications for tenure (the award of permanent job status by research and academic institutions) and applications for promotion by tenured scientists.

Summarizing Scientific Literature

Once the relevant literature has been located and assembled, the scientist must read, understand and summarize one to many publications and/or other documents so that meaningful comparisons of studies conducted by different researchers may be made. The researcher should read each article one or more times to ensure that he or she understands the author's message, and then critique or summarize the article (or sections thereof) **in his or her own words**. Any concerns or comments regarding the objectives, methodology and interpretation of results should be clearly noted. One effective way to accomplish this is to create a word-processing document with bibliographic information (author(s), year, title, journal name, volume number, and page numbers) for each manuscript reviewed. One or more sections of each manuscript may be then "cut and pasted" into the file (verbatim) followed by the researcher's or reviewer's comments (in his or her own words). Under no conditions should the researcher summarizing the manuscript use the original author's wording (or anything close to it) in his summary as many readers may interpret this as plagiarism, a serious form of academic misconduct. When a series of such summaries have been completed, a comparison of similarities and differences in interpretation of research conducted by different researchers under different conditions should be relatively straightforward, but must be composed in the researcher's own words.

Example 2.2—Summarizing a Manuscript

MANUSCRIPT BY SMITH ET AL. (2014) J. Sci. 41:50–65. Methodology, p.55.

Original text: Cohorts of white rats were maintained in cages in the laboratory under a constant temperature of 20 (±1)°C, 70% humidity (±1), and a 12L:12D photoperiod. Rats were divided into 3 groups of 10 each and were fed 100g, 150g, or 200g of diet daily (dependent on group they were assigned to) and were provided an unlimited amount of fresh (tap) water. Initial weights of rats in each group were measured at the beginning of the experiment and final weights were measured at the conclusion.

Reviewer's Comments: There appear to be some major problems in the way in which results were reported: (1) the author failed to identify the variety, age, and sex of the rats used in the experiment, (2) the author did not discuss the type of cage (wire, wood, etc.) that rats were maintained in, or how many were enclosed in each, (3) author failed to discuss the rationale for a 12L:12D photoperiod rather than some other combination, 4) author failed to mention the sensitivity of the weighing instrument and the time of day weights were measured. Note: because of these limitations, his conclusion that rats fed his diet responded differently than those in Spencer's study do not appear to be justifiable.

The method discussed above is a variation of the "split-page" method of note-taking discussed in Pechenik (2013) and other textbooks on scientific writing. Notice that the researcher has pointed out several concerns regarding the methodology of this research, and has suggested that these limitations may not justify the original author's conclusions regarding an earlier study by Spencer. Summarizing publications (or sections thereof) in this manner greatly facilitates comparisons of research conducted by different authors on similar topics, and provide considerable insurance against plagiarizing other scientists (plagiarism and other forms of academic misconduct are discussed in greater detail in Chapter 5).

Citing Scientific Literature

All scientific literature referenced in written reports and manuscripts must be properly cited in both the text and a references section of the document using an approved citation format. Several citation formats are in common use (Table 2.3), and the format required by a particular journal may be determined by consulting the *Instructions to Authors* (or *Contributors*) statement posted on the journal's website and/or by examining the *Literature* (or *References) Cited* section of an article published in a recent issue of the journal.

Table 2.3 Style manuals in common use for scientific and other publications.

Style Manual	Web Site
MLA Handbook for Writers of Research Papers, 7th edition, 2009–10.	www.mlahandbook.org
The Chicago Manual of Style, 16th edition, 2010.	www.chicagomanualofstyle.org
Scientific Style and Format: The CSE Manual for Authors, Editors, and Publishers, 8th edition, 2014.	www.scientificstyleandformat.org
Publication Manual of the American Psychological Association, 6th edition, 2010.	www.apastyle.org/
American Medical Association Manual of Style, 10th edition, 2015	www.amamanualofstyle.com/
Turabian—A Manual for Writers of Term Papers, Theses, and Dissertations, 6th edition, 2013	www.press.uchicago.edu/

The References section of this book follows the *Chicago Manual of Style* (www.chicagomanualofstyle.org/), while the following examples follow the formats of *The Scientific Style and Format:* CSE Manual for Authors, Editors and Publishers, 8th edition, which is used most commonly in scientific publications (www.councilscienceeditors.org/). In this system, the format used for a particular article is based on the type of article (e.g., journal vs. book or book chapter) and the number of authors listed:

- **journal article—single author**

  If citation occurs at the beginning of a sentence or is embedded within a sentence, list the author's last name followed by year of publication enclosed in parenthesis (a). If citation occurs at end of a sentence, enclose entire citation in parentheses (b).

  (a) Smith (1960) if embedded or located at beginning of sentence.
  (b) (Smith 1960) if located at end of sentence.

  In the *References Cited* section, list the author's last name followed by initials, date of publication, title of article, journal name, volume, and page numbers. In journal articles (and books), only the first major word of title is capitalized— the remainder should be in lower case, and the title is not enclosed within parentheses. Note also that journal names are commonly abbreviated. For example,

  Smith, R I. 1960. Title of article. J. Hypothetical Res. 10:100–125.

  If an author publishes more than one journal article in the same year (as senior author), the first mentioned article should be designated as "a," the second mentioned as "b," and so on. In the References Cited section, these designations should be attached to the year of publication. For example,

  (a) Smith (1965a) and Smith (1965b) in text.
  (b) Smith, RJ. 1965a. Title of article. J. Hypothetical Res. 25:1–8.
  (c) Smith, RJ. 1965b. Title of article. J. Hypothetical Res. 25:55–60.

- **journal article—two authors**

  Format for citations in text is similar to that for journal articles—single author, except that both author's last names are listed followed by date of publication in parentheses according to rules (a) and (b) discussed previously. For example,

  (a) Smith and Jones (1962) if embedded or located at beginning of sentence.
  (b) (Smith and Jones 1962) if located at end of sentence.

  Format for citations in *References Cited* section is similar to that for journal articles—single author, except that both authors names are listed as follows:

  Smith, RI, and Jones, BJ. 1962. Title of article. Hypothetical Ecol. 42:25–28.

- **journal article—three or more authors**

  Format for citations in text is similar to that for journal articles—single author, except that the first author's name is followed by *et al.* (Latin for "and others") and the date of publication as described previously. For example,

  (a) Smith *et al.* (1965) if embedded or located at beginning of sentence.
  (b) (Smith *et al.* 1965) if located at end of sentence.

  Format for citations in *References Cited* section is similar to that for journal articles—single author, except that both authors names are listed as follows:

  Smith, RI, Jones, BJ, Gutierrez, JM. 1965. Title of article. J. Hypothetical Ecol. 42:59–68.

- **book—single author and/or multiple authors**

  Format for citations in text is similar to that of journal article—single author and others discussed previously. Citations listed in *References Cited* section should include author(s) names followed by initials, date of publication, title of book with all major words capitalized, location (state) of publisher, name of the publisher, extent in terms of pages (optional). For example,

  Smith, RI, Jones, BJ, Gutierrez, JM. 1965. Hypothetical Ecology. Science Publishers, New York. 550 pp.

- **chapter in edited book—single and/or multiple authors**

  Format for citations in text is similar to that of journal article—single author and others discussed previously. Citations listed in *References Cited* section should include author(s) names followed by initials, date of publication, title of article with first major word capitalized and all others in lower case, title of book with all major words capitalized, name of the publisher, location, and the total number of pages in the book. For example,

  Smith, RI, Jones, BJ, Gutierrez, JM. 1965. Hypothetical Theory, 2nd ed. Science Publishing, New York. 755 pp.

- **thesis or dissertation**

  Format for citations in text is similar to that of journal article — single author and others discussed previously. Citations listed in *References Cited* section should include author's name followed by initials, date of approval, title of thesis or dissertation with first major word capitalized and remaining words in lower case, followed by a designation of "thesis" or "dissertation" in brackets, name of the university, and extent of document in terms of page numbers. For example,

Jones, BJ. 1966. Transmission of plant pathogens by cotton aphids [dissertation]. Hypothetical University. 125 p.

- **web site**

Format for citations in text is similar to that of journal article—single author and others discussed previously. Citations listed in *References Cited* section should include author(s) names followed by initials, date of posting, title of article with first major word capitalized and all others in lower case followed by the designation "Internet" in brackets, location and name of organization, date cited, and the heading "Available from" followed by the source UTL. For example,

Smith, RN. 2005. New fish species in Tennessee caves (Internet). Nashville (TN): University of Tennessee: [cited 2012 Oct 14]. Available from http://www.phoneysite.org.

- **personal communication**

Unpublished information obtained from colleagues and discussed in text (with colleague's knowledge and permission) should be cited in text as follows:

(Dr. Ronald Davids, USDA, personal communication, 2005).

Note: Before citing a personal communication in a presentation or publication, make certain that you have the approval of the person cited in writing—otherwise you may encounter serious legal problems (see Chapter 13).

- **citing an article that was not actually read**

The CSE Manual does not currently recommend the citation of such secondary sources of information and suggests that the author locate the original source document. If the original document cannot be located and is considered important enough to cite, use the format recommended in Pechenik (2013):

(Smith 1904, as cited by Jones 2011) in text.
Cite both references in the Literature Cited section.

Exercise 2.1

Using one or more of the online databases, try to locate the book author's M. S. thesis on biotype-C greenbug and his Ph.D. dissertation on citrus blackfly. Do the same for the instructor in your course.

Title of M.S. thesis ______________________________________
Date approved ____________________ by University ____________________
Title of Ph.D. dissertation ______________________________________
Date approved ____________________ by University ____________________
____

References

Pechenik, J. A. (2013). *A short guide to writing about biology.* New York: Pearson, 276 pp.

Test Your Knowledge

Without referring to the material presented earlier in this chapter, mark an "x" in the box of the most appropriate answer to each question. Answers are provided at end of chapter.

1. In general, literature searches for a proposed project involving field sampling should be conducted ______________.
   - ❑ **a.** before the project is initiated in order to identify gaps in knowledge,
   - ❑ **b.** at the time the project begins to ensure that the proposed sampling protocol is appropriate,
   - ❑ **c.** while the project is in progress to ensure that any new literature published after the project began is identified,
   - ❑ **d.** after the project has terminated in order to ensure that all relevant literature on the research topic has been identified and studied.

2. **True or False**. A typical *online database* provides access to literature contained in its own database, which may not include all sources of primary and secondary literature.
   - ❑ **a.** True
   - ❑ **b.** False

3. In order to obtain a report of literature sources dealing with mosquitoes as vectors of malaria, which of the following would you enter in an online database?
   - ❑ **a.** mosquitoes AND malaria
   - ❑ **b.** mosquitoes OR malaria
   - ❑ **c.** mosquitoes NOR malaria
   - ❑ **d.** mosquitoes XOR malaria

4. If you substituted OR for AND in the previous command (question #3), the number of literature sources reported would almost certainly ______________.
   - ❑ **a.** increase
   - ❑ **b.** remain the same
   - ❑ **c.** decrease
   - ❑ **d.** cannot be determined from this information alone

5. The terms AND, OR, NOR and XOR are commonly referred to as ______________.
   - ❑ **a.** keywords
   - ❑ **b.** Search Operators
   - ❑ **c.** Boolean Operators
   - ❑ **d.** Algebraic Indicators

6. True or False. Using identical search commands, search engines will almost invariably report a larger number of 'hits' than online databases.

❑ **a.** True

❑ **b.** False

7. The reason for the discrepancy discussed in the previous question (#6) is that ______________________________________.

❑ **a.** search engines are inherently more efficient than online databases

❑ **b.** online databases do not generally report sources with imagery

❑ **c.** online databases are less compatible with Internet protocol than search engines

❑ **d.** search engines cover not only literature sources, but also web sites, blogs and other sites that are outside the realm of online databases

8. ResearchGate®, LinkedIn®, and Academia.edu® are examples of ____________________________.

❑ **a.** social media for scientists

❑ **b.** search engines

❑ **c.** blogs for scientists

❑ **d.** web sites for refereed journals

9. In literature citations, the Latin term *et al.* is used to indicate ___________________.

❑ **a.** that the reference was cited earlier in the text

❑ **b.** that the article was written by an anonymous author

❑ **c.** that three or more individuals are included in author list

❑ **d.** that the article is "in press" (not published yet)

10. *True or False.* It is not considered appropriate to cite online web sites in scientific articles and documents.

❑ **a.** True

❑ **b.** False

Answers: 1) a; 2) a; 3) a; 4) a; 5) c; 6) a; 7) d; 8) a; 9) c; 10) b.

Designing, Conducting, and Interpreting Surveys and Experiments

The Essence of Science

Chapter Learning Objectives

After studying this chapter, you should understand:

- The scientific method and its role in the explanation of natural phenomena.
- The concept of sampling and estimating population parameters.
- The design, statistical analysis, and interpretation of simple experiments.
- Important factors to consider when reporting results of statistical analyses in the scientific literature.
- The importance of maintaining laboratory and field notebooks when conducting scientific research.

One does not have to be a statistician in order to design and conduct meaningful scientific research, but it is necessary for a scientist to understand some basic statistical concepts and terminology in order to design and conduct valid experiments and to report the results in the scientific literature. A basic knowledge of statistical concepts and terminology is also a requisite for other activities that scientists routinely engage in (e.g., providing anonymous manuscript reviews for journal editors and reviewing grant proposals submitted by other scientists to funding agencies). Thus, the objective of this chapter is to provide an overview of basic statistical concepts and terminology that all scientists should be familiar with, and will begin with a definition of the scientific method.

The Scientific Method

The *scientific method* has been used by scientists for several centuries to evaluate and explain natural phenomena in a rational manner using a combination of observations and the formulation of tentative explanations (*hypotheses*) which can be tested empirically in experiments (Gower, 2002). It was elucidated by Rene Descartes, Francis Bacon and others during the Scientific Revolution in 17th-century Europe and continues to form the cornerstone of modern science. In a typical situation, repeated observations of a phenomenon may lead to a generalization through *inductive reasoning* (e.g., plants growing near piles of animal excrement are usually larger and healthier than plants of the same species growing in areas of the same field devoid of excrement). Generalizations such as this commonly lead to the development of hypotheses through *deductive reasoning* (in this case, that animal excrement contains one or more substances that facilitate plant growth and health) that can be tested experimentally.

An *experiment* is designed to evaluate the magnitude of changes that occur in a *response* or *dependent* variable (e.g., growth rate and/or average yield of a certain crop species) as values of an *explanatory* or *independent* variable are manipulated or varied (e.g., levels of an experimental fertilizer or plant growth hormone) while all other variables affecting the response variable are held constant or occur at similar levels within all experimental groups. In this respect, experiments differ fundamentally from *surveys* and *studies,* in which variables are measured but no attempt is made to manipulate levels of any of them. By analyzing measurements of the dependent variable collected during the course of an experiment using appropriate statistical tests, the effects of the independent variable can be evaluated, and the probability of these results being repeatable (a requisite for valid experiments) can be determined. Use of well-designed experiments is now universally accepted by scientists and can greatly minimize or eliminate spurious results in research caused by various biases (conscious and unconscious) on the part of the researcher.

Designing and Conducting Surveys and Experiments

Several important considerations must be addressed when designing and conducting surveys and experiments:

- the response and independent variables must be clearly defined and measured as accurately as possible,

- in contrast to surveys, experiments must incorporate both *treatment(s)* and a *control*, e.g., in the example discussed in the previous section, the treatment(s) would be groups of plants exposed to various levels of animal excrement and the control would be plants growing in the absence of excrement, with all other variables the same in both groups,

- the statistical test that is to be used in analyzing data should be chosen before the experiment begins, (i.e., don't collect data first and then decide how you are going to analyze it),

- sample sizes should be adequate for the purposes of the experiment (in general, "large" samples are those with ≥30 observations, while "small" samples consist of fewer observations),

- all procedures and data should be maintained in a field or laboratory notebook,

- data that appears to be unusual or "bad" should never be deleted from data sets unless justified on statistical grounds (tests for outliers exist),

- when interpreting results of statistical tests, remember that statistics do not "prove" anything. Rejection of the *null hypothesis* (i.e., that no difference exists among the means) should be interpreted as ". . . our results indicated significant differences among the means . . ." or something equivalent. Failure to reject the null hypothesis should be reported as ". . . our results failed to detect significant differences among the means . . ." or something equivalent. In other words, the interpretation of experimental results is based solely on results of statistical tests, not on opinions of the investigator(s).

Estimating and Comparing Population Parameters

In the scientific world, the term *population* has several meanings. A *biological* population is defined as members of the same species inhabiting a certain geographic area during a certain time interval. On the other hand, a *statistical* population is usually defined as members of a group of objects (biotic or abiotic) with certain characteristics that can be measured and compared with members of other populations. For example, groups of students in a university (biotic) and groups of stones in a geographic location (abiotic) are both statistical populations with characteristics (e.g., size and weight) that can be measured and compared with similar groups from other locations. Each group is also characterized by certain population *parameters* (e.g., the average size and average weight of all individuals in the population at a given time) which exist but may be impossible to measure with 100% accuracy because of limitations of measuring instruments and other factors. If a population consists of only a few individuals, it may be possible to measure all individuals and obtain estimates of parameters that are very close to their actual values (this would be analogous to a *census*). If the population is large, however, this method of estimation is not feasible and a procedure known as *sampling* must be used.

Sampling involves the selection and measurement of subgroups of individuals that are presumed to be representative members of the population under study. In order to be representative, samples must be unbiased (i.e., the procedure used must not intentionally or unintentionally exclude any segment of the population) and must be selected randomly (i.e., the selection of one individual should not influence the selection of any other) in numbers sufficient to estimate variation occurring within the population (in general, a sample size of 30 or greater is considered a "large" sample). If these conditions are met, the following *sample statistics* provide unbiased estimates of the corresponding *population parameters* with a known level of confidence:

- *mean* ($\overline{x}$), is defined as

$$\overline{x} = \sum_{i=1}^{n} n_i / N$$

where n_i = the value of individual observations and N = sample size. This is the familiar sample mean, which is an unbiased estimate of the population mean (μ).

- *variance* (s^2), is defined as

$$s^2 = \sum_{i=1}^{n}(x_i - \bar{x})^2 / N-1$$

where x_i = the value of individual observations and $\bar{x}$ = the sample mean. This is an unbiased estimate of the population variance (σ^2).

- *standard deviation* (s), is defined as

$$s = \sqrt{s^2}$$

where s^2 = the sample variance. The sample standard deviation is an unbiased estimate of the population standard deviation (σ). In a standardized normal distribution (i.e., with mean 0 and standard deviation (SD) of 1), 68.3% of the data lie within ±1 SDs of the mean, 95.4% lie between ±2 SDs of the mean, and >99% lie between ±3 SDs of the mean (Figure 3.1).

- *standard error* ($s_{\bar{x}}$), is defined as

$$s_{\bar{x}} = s / \sqrt{n}$$

where s = the sample standard deviation. This statistic is also known as the *standard deviation of means*, and is used to calculate another important estimator of dispersion.

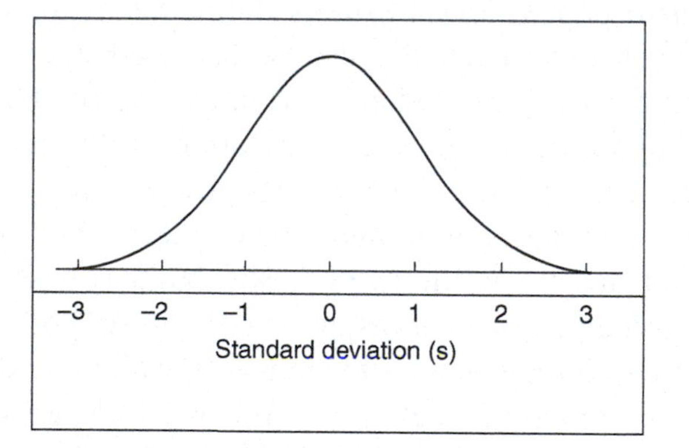

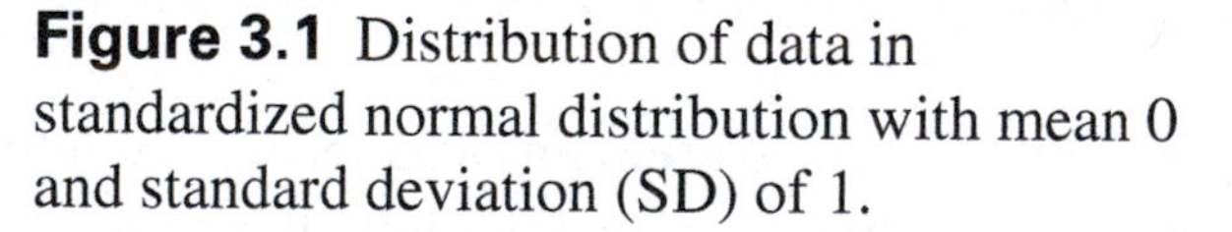

Figure 3.1 Distribution of data in standardized normal distribution with mean 0 and standard deviation (SD) of 1.

Confidence intervals are regions defined by upper and lower *confidence limits* that are expected to include a certain percentage of the means of future samples based on similar sampling procedures and similar sample sizes. For example, a 95% confidence interval is defined as

$$CI = x \pm t_{.05}\, s_{\bar{x}}$$

where x is the sample mean, $s_{\bar{x}}$ is the standard error, and $t_{.05}$ a critical value from the t-distruibution (1.96 for 95% confidence limits). Thus, if a future study were conducted using the same sampling procedures and similar sample sizes, approximately 95% of the sample means calculated would fall within the 95% confidence intervals. Another interpretation is that the 95% confidence intervals would include the true (but unknown) mean of the population 95% of the time. Because of its importance in scientific research, sampling theory is covered extensively in the scientific literature and continues to be an active area of research (Sokal & Rohlf, 1995).

Example 3.1 – Confidence Intervals

The measurements below represent body lengths (snout to venter) of adult female lizards of a certain species inhabiting an island located in a coastal zone of the eastern United States.

Body Length (cm)	Summary Statistics
4.1	n = 5
4.3	$\bar{x}$ = 23.1 / 5 = 4.62
5.0	s2 = [(4.6 – 4.1)2 + . . . + (4.6 – 5.1)2] / 4 = 0.187
4.6	s = √ 0.187 = 0.432
5.1	$s_{\bar{x}}$ = .432/2.236 = 0.193
23.1	

95% Confidence Intervals

$$CI_{.95} = 4.62 \text{ +/– } 1.96\ (0.193) = 4.62 \pm 0.378 = (4.242,\ 4.998)$$

Based on this particular sample, there is a 95% probability that the true population mean of this population lies between 4.242 and 4.990. If this same population was sampled repeatedly in a future study using the same procedures and same sample sizes, the sample means and width of the 95% confidence intervals would be expected to vary from sample to sample. However, the calculated 95% confidence intervals for individual samples would be expected to include the true (but unknown) population mean 95% of the time.

Comparisons of Sample Estimates—Two Means

Confidence intervals provide a simple way to evaluate whether or not two sample means are members of the same population. This comparison is based on a null hypothesis that the two means are similar, that is, H_o: $\bar{x}_1 = \bar{x}_2$, which is accepted if the 95% CIs overlap and rejected if the 95% CIs do not overlap. In the examples below, the CIs for means in the first pair overlap and hence, the null hypothesis is not rejected (a) whereas the CIs in the second pair do not overlap and hence, the null hypothesis is rejected, which implies that the means are different at the 5% probability level. The term *significance* in this case means that if the same sampling procedures and sample sizes were used in future experiments, the true means of the populations would fall within the CIs 95% of the time.

a. 4.3--------5.2---------6.1 — CIs overlap—Accept null hypothesis

b. 4.2--4.6---5.0

c. 1.6--2.3--3.0 — No overlap—Reject null hypothesis

d. 4.2--4.6--5.0

Two other tests are commonly used to compare means from sample populations that are assumed to be normally distributed (see Figure 3.1).

- **Z-test** is used if sample size is large ($n > 30$ observations).
- **Student's *t*-test** is used if sample size is small (<30 observations). Several forms of the Student's *t*-test exist, and the appropriate form for a particular dataset depends on whether the samples are independent or paired, and whether the variances are equal or unequal.

Example 3.2 – Student's t-test

Blood_ Glucose_ Level

Group 1	Group 2
110.25	200.1
105.75	215.15
103.35	201.35
101.2	210.5
97.05	208.6
100.1	206.55
102.15	207.95
104.2	225
98.3	223.2
107.55	220.4
106.6	224.25
105.75	209.35
100.8	208.65
103.25	200.1
102	215.05

Summary Statistics

	Group 1	Group 2
$\bar{x}$	103.2	211.7
s^2	12.7	72.0
s	3.5	8.5
$s_{\bar{x}}$	0.9	2.2
95% CL	±1.97	±4.70

F test for Variances
$F = 0.145$; $df = 13,13$; $P < 0.001$
(critical $F = 0.388$)

t-test 2-sample unequal variances
$t = 47.156$; $df = 17$; $P < 0.001$
(critical $t = 2.10$)

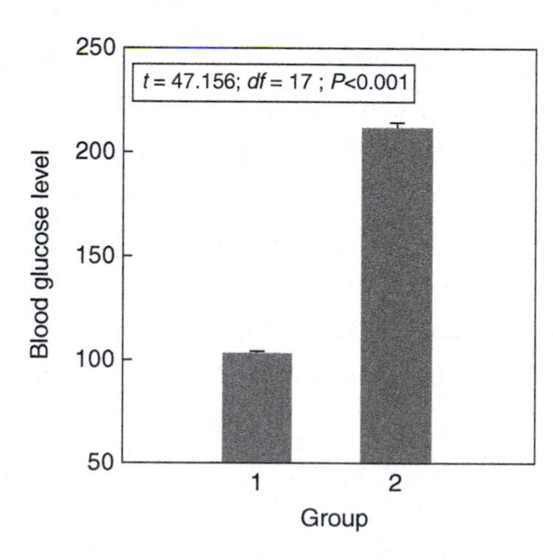

Figure 3.2 Bar chart of data for blood glucose levels indicating differences in means between two treatment groups. Error bars are included to provide reader with information on variability in samples, and statistics for the two-sample *t* test assuming unequal variances indicate that the means of the two groups are significantly different at the 5% probability level.

In this example, variances of blood glucose levels in the two groups were unequal ($F = 0.145$; $df = 13{,}13$, $P < 0.001$ units) and the appropriate t test indicated that the mean for group 1 (103.2 units) was significantly higher than group 2 (211.7) ($t = 47.2$; $df = 17$; $P < 0.001$). The P-value represents the probability of a Type 1 error (i.e., rejecting the null hypothesis of no difference when a real difference actually exists) is <0.001 and implies that if this study were repeated 100 times using the same sample size and sampling methodology, we could expect to arrive at the same conclusion 99% of the time (1% would be different due to random error). Results of statistical tests may be incorporated into the text of a report or in figures (Figure 3.2).

Comparison of Sample Estimates—3 or More Means

When means from three or more normally distributed sample populations are to be compared, an *analysis of variance* (ANOVA) is generally used. The type of ANOVA that is appropriate for a particular dataset is dependent on the experimental design, and the complexity ranges from simple to complex. Common types of ANOVA include the following:

- **Completely randomized design**. This is one of the simplest types of ANOVA and involves a single response variable (in the case below, yield) and a single independent variable (in the case below, soil type). The assumptions are that the samples are independent and randomly selected from populations that are normally distributed with equal variances.

Example 3.3 – ANOVA – Completely Randomized Design

The data below represent yields of a certain crop cultivar (the dependent variable) grown in three soil types (the independent variable).

	Yield (kg/row)		
	Sand	Silt	Clay
	50.0	67	45.0
	55.0	69	44.0
	54.0	65	43.0
	51.5	70	46.0
Mean	52.6	67.8	44. 5

A completely randomized (or one-way) ANOVA detected significant differences among the means for plants grown in sand (52.6 kg/row), silt (67.8 kg/row) and clay (44.5 kg/row) ($F = 101.85$; $df = 2{,}6$; $P < 0.001$). These results indicate that at least two of the means differ, but the ANOVA table alone does not identify specifically where the differences occur. For this purpose, a means comparison test (the Tukey's HSD in this case) was used to compare sand vs silt ($P < 0.001$), sand vs clay

ANOVA

Source of Variation	*SS*	*df*	*MS*	*F*	*P*-value	(Critical F)
Between Groups	854.39	2	427.19	101.85	<0.001	5.14
Within Groups	25.17	6	4.19			
Total	879.56	8				

($P < 0.0001$) and silt vs clay ($P < 0.0001$). In all three cases, the P-values were less than 0.05 and the null hypothesis of no differences among means was rejected. The interpretation of this comparison is that the lowest yields of this particular cultivar occurred when plants were grown in clay soil, the next highest yields occurred when plants were grown in sandy soil, and the highest yields occurred with plants that were grown in silt. Results such as these can be summarized in the form of bar charts (Figure 3.3) or in tables (Table 3.1). When reported in tabular form, differences among means are commonly denoted by letters with explanatory footnotes identifying the statistical test used and probability level (Table 3.1).

- **Randomized complete block design.** A second common form of ANOVA is the randomized complete block design (also known as a two-way ANOVA), which is used to evaluate the effects of two factors (independent variables) on the response (dependent) variable.

Means Comparisons (Tukey HSD Multiple Comparisons)

	Sand	Silt	Clay
Sand	1.000		
Silt	0.001	1.000	
Clay	0.000	0.000	1.000

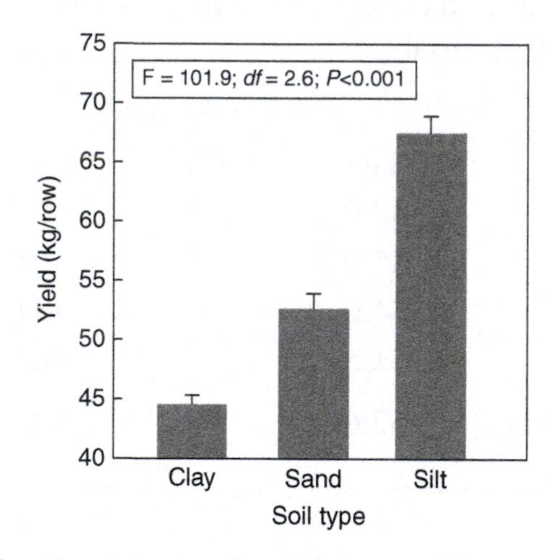

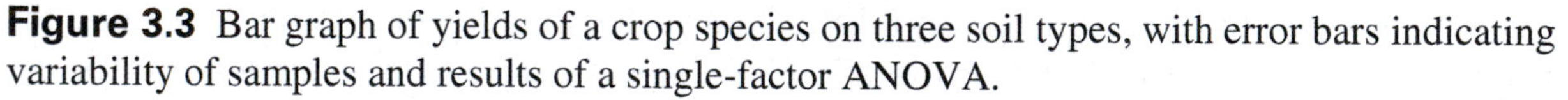

Figure 3.3 Bar graph of yields of a crop species on three soil types, with error bars indicating variability of samples and results of a single-factor ANOVA.

Table 3.1. Comparison of crop yields occurring on three soil types in southern Texas.

Soil	Mean ± SE[1]
Sand	52.6 ± 1.14a
Silt	67.8 ± 1.11b
Clay	44.5 ± 0.65c

[1]Means followed by same letter are not significantly different at 5% probability level (Tukey's HSD test).

Example 3.4 – ANOVA – Randomized Complete Block design

The data below represent yields of the crop cultivar discussed in Example 3.3 (dependent or response variable) grown in a series of four fields with silty soil (factor 1—locations or blocks) subjected to three rates of nitrogen fertilizer (factor 2). The two-way ANOVA detected significant differences among the means for fertilizer rate ($F = 75.0$; $df = 2,2$; P = 0.013) and no differences among locations (blocks).

	Fertilizer Rate (kg/row)		
	Low	Medium	High
	45	62	75
	44	61	71
	43	64	76
	46	66	74
Mean	44.3	63.7	73.7

ANOVA

Source of Variation	*SS*	*df*	*MS*	*F*	*P*-value	(Critical F)
Rows	21.3	2	10.67	5.33	0.157	19
Columns	150.0	1	150.0	75.0	0.013	18.5
Total	175.3	5				

Means Comparisons (Tukey HSD Multiple Comparisons)

	Low	Silt	Clay
Sand	1.000		
Silt	0.001	1.000	
Clay	0.000	0.000	1.000

Table 3.2 Comparison of crop yields occurring in three sites subjected to three rates of N fertilizer.

Rate (kg/ha)	Mean ± SE[1]
Low	44.3 ± 0.88a
Medium	63.7 ± 1.45b
High	73.7 ± 1.45c

[1]Means followed by same letter are not significantly different at 5% probability level (Tukey's HSD test).

- Other common experimental designs that are analyzed using ANOVAs are summarized in many of the references listed at the end of this chapter.

Correlation and Regression Analysis. Two related statistical procedures—correlation and regression analysis—are commonly used to evaluate the degree of association between two or more variables, that is, how changes in one variable are associated with changes in a second variable. Correlation coefficients measure the strength of the relationship (if one exists) and indicate whether the correlation is *positive* (if *X* increases, *Y* increases accordingly) or *negative* (if *X* decreases, *Y* decreases accordingly). It is important to note that while significant correlation coefficients provide evidence of associations between variables, they do not imply cause-effect relationships. For example, height and weight in humans tend to be positively correlated, but height *per se* is not the principal determinant of a person's weight (numerous other physiological and behavioral factors are involved). In contrast, regression analysis provides a mathematical means by which to predict the value of a response variable in relation to changes in an independent variable, and may be used to evaluate possible cause-effect relationships. For example, the effect(s) of calorie intake (an independent variable) on weight gain in humans (a response or dependent variable) may be evaluated experimentally by providing groups of sedentary (nonexercising) humans with diets containing known quantities of calories, and measuring changes in weight for each subject in each group during a specified time period. By regressing values of the dependent variable on the independent variable, an equation indicating the intercept and slope for a best-fitting line through the data is obtained from which predictions of weight gain as a function of calorie intake may be made (procedures are discussed below). When using regression models, it is important that predictions are valid only within the range of the data and extrapolations outside the data range should not be attempted. For example, a regression model of insect developmental rates (response variable) as a function of ambient temperatures (independent variable) may be accurate within the range of temperatures that were used to develop the model, but meaningless outside this range where development is impeded or prevented.

Nonparametric Statistics. When the assumptions underlying parametric tests are not satisfied, several distribution-free (nonparametric) tests are commonly used to test whether or not two or more populations have the same distribution or location (Stephens, 2004).

- Sign test—used with ordinal (ordered) data or when assumptions for Student's *t*-test are not met
- Wilcoxon Rank Sum Test—used when independent samples *t* test are not satisfied
- Wilcoxon Signed Rank Test—used for Paired Difference experiment
- Kruskal-Wallis test—used when assumptions for completely randomized ANOVA are not met
- Friedman test—used when assumptions for randomized block ANOVA are not met
- Spearman Rank Correlation Coefficient—used to provide measure of correlation between two sets of ranks

Example 3.4 – Correlation and Regression Analyses

The data below represent values of an as independent variable (X) and response variables for 3 different populations (Y1, Y2, Y3) (Figure 3.4a). Scatter plots and correlation coefficients indicated a strong positive correlation between X and Y1 (Figure 3.4b), no correlation between X and Y2 (not shown), and a strong negative correlation between X and Y3 (Figure 3.4c). A regression of Y1 on X was significant ($F = 208.07$; $df = 1,9$; $P < 0.001$) and produced a trend line with equation $Y = 4.15 + 0.88X$ (Figure 3.5). The first term represents the predicted Y-intercept while the second term represents the slope, that is, the magnitude of change in Y1 that is predicted to occur for each unit change in X (in this case, a unit change in X is predicted to result in a 0.88 change in Y1). The coefficient of determination ($R^2 = 0.9585$) indicates that approximately 96% of the variation in the data is explained by the regression.

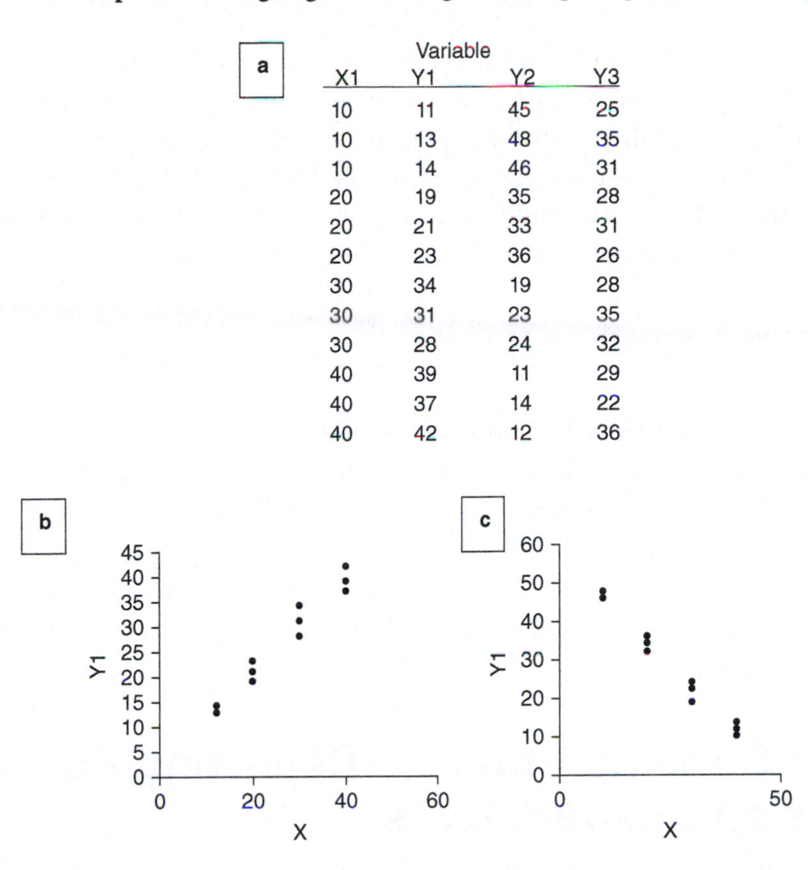

a

	Variable		
X1	Y1	Y2	Y3
10	11	45	25
10	13	48	35
10	14	46	31
20	19	35	28
20	21	33	31
20	23	36	26
30	34	19	28
30	31	23	35
30	28	24	32
40	39	11	29
40	37	14	22
40	42	12	36

Figure 3.4 Data for an independent variable (*X*) and three response variables (*Y*1–*Y*3) from different populations (a). Scatter plots and correlation coefficients indicating a strong positive correlation ($r = 1.0$; Figure 3.3b) and a strong negative correlation ($r = -1.0$; Figure 3.3c). Note that positive correlations are indicated by positive (+) signs and negative correlations are indicated by negative (–) signs.

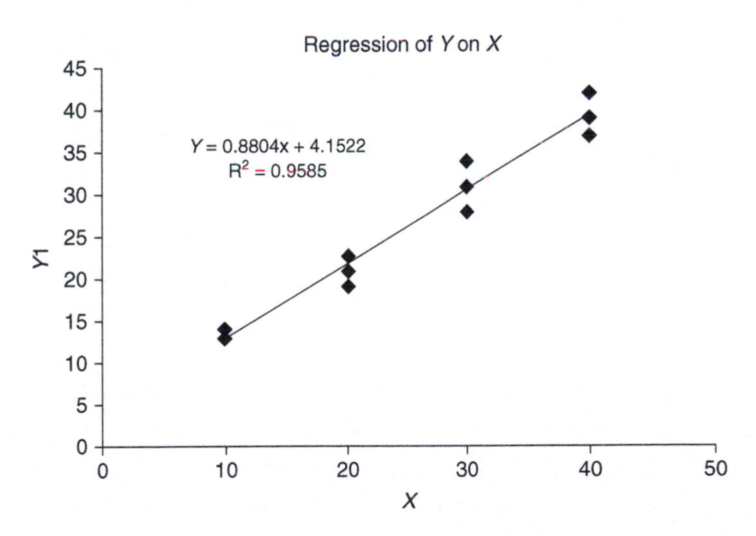

Figure 3.5. Regression of *Y*1 on *X* produced a trendline indicating a y-intercept of 4.15 and a slope of 0.88*X*. The coefficient of determination indicates that approximately 96% of the variation of the data is explained by the regression.

Chi Square tests are used to test a variety of hypotheses involving categorical and other types of data. One simple and common example is to compare frequencies of mortality occurring in the presence and absence of a mortality factor. Suppose that, 200 individuals in a population were exposed to and protected from a predator and counts of live and dead individuals were counted at the conclusion of the experiment. Of the 100 individuals exposed to the predator, 90 died during the course of the experiment, compared to only 50 which died in the protected cohort. The chi-square criterion is calculated as follows

$$\chi^2 = \Sigma \text{ (observed} - \text{expected)}^2 / \text{expected}$$

with $n - 1$ degrees of freedom (calculation of observed and expected values for chi square are discussed in Stephens [2004] and other references at the end of this chapter). In this case, the χ^2 value of 38.02 was larger than the tabular χ^2 (11.34) and the null hypothesis was rejected ($P < 0.01$). The interpretation of the experiment was that the cohort exposed to predators suffered a substantially higher mortality than contemporaries protected from predators but exposed to other elements that affected both cohorts. Other examples of chi-square tests in statistical applications are provided in references at the end of this chapter.

Important Considerations in Reporting Results of Studies and Experiments

Reporting the results of scientific studies and experiments involves a writing style that differs substantially from other types of writing. Statistical comparisons and tests are based on probabilities and are almost always based on a null hypothesis that the "treatments" evaluated do not differ from each other. Thus, when a researcher designs and conducts an experiment to compare the effects of fertilizer rates on plant growth, he or she usually includes two or three fertilizer rates (including a control with no fertilizer) and compares the means of each treatment using an ANOVA. The null hypothesis in this case would be that the fertilizer has no effect on plant growth (even though the researcher may not believe this to be true), and would be rejected only if statistical evidence suggests that it is not correct.

When reporting results of experiments, several factors must be kept in mind:

- the experimental design must be explained thoroughly and clearly in the materials and methods sections of manuscripts,

- remember that statistical tests are based on probabilities and null hypotheses assume that no differences among treatment means exist—the null hypothesis is rejected only if justified by the statistical test used,

- never delete "bad" data points arbitrarily unless justified on statistical grounds—unusual data points that aren't consistent with the majority of data points collected may be due to faulty equipment or collection techniques (in which case deletion is clearly justifiable) but may also be the result of important natural factors the researcher is unaware of (in which case deletion would confound interpretation of the data),

- when reporting statistical results, always provide the reader with information on variability of data and results of *t*-tests, ANOVAs, and other analyses in either figures, tables, or in the text of manuscripts.

Importance of Laboratory and Field Notebooks

The importance of maintaining laboratory and field notebooks when conducting scientific research cannot be overemphasized. Accurate notebooks are indispensable when summarizing experimental procedures for publications (which is normally done after research has been completed) and for documenting events which may have influenced results of one or more experiments (e.g., an unexpected disease outbreak in cultures of plants used in growth comparisons). Laboratory notebooks are absolutely critical in research and development that involves proprietary products and possible patents. Human memories are fallible—**do not try to remember what occurred during an experiment!**

Guidelines for creating and maintaining laboratory and field notebooks include the following:

- notebooks should be bound and paginated,
- entries should be made using blue or black ink, and should be dated and signed or initialed by the investigator (in the case of research involving the development of proprietary products, the signature(s) of one or more witnesses should be included),
- experimental procedures should be described in detail, and any ancillary material (e.g., imagery or maps) should be stapled to the appropriate page,
- any event(s) which may have influenced the experiment should be documented clearly at the time it occurred,
- archive the notebook(s) after the research has been completed and after any patent(s) have been awarded.

Exercise 3.1

Using the data provided in Exercise 3.4, run a correlation analysis between *X* and *Y*2 and between *X* and *Y*3 at the 5% probability level using Microsoft® Excel (ask your instructor for help if you are unfamiliar with EXCEL). Interpret the correlation coefficient and explain what it means.

References

Gower, B. (2002). *Scientific method: A historical and philosophical introduction* New York: Taylor & Francis e-Library, Routledge, 276 pp.

Rohlf, F. J., & Sokal, R. R. (1995). *Statistical tables* (3rd ed.). New York: W. H. Freeman and Company, 199 pp.

Sokal, R. R., & Rohlf, F. J. (2012). *Biometry: The principles and practice of statistics in biological research* (4th ed.). New York: W. H. Freeman and Company, 915 pp.

Staff. (2003). *Statistics. Research and education association* NJ: Piscataway, 318 pp.

Staff. "Lab Notebooks," http://techtransfer.tufts.edu/resources/tufts-policies/lab-notebooks/ (February 22, 2016)

Stephens, L. J. (2004). *Advanced statistics demystified*. New York: McGraw-Hill, 324 pp.

Wikipedia, "History of the Scientific Method," *Online,* http://en.wikipedia.org/wiki/historyof scientificmethod (April 2, 2015)

Test Your Knowledge

Without referring to the material presented earlier in this chapter, mark an "x" in the box of the most appropriate answer to each question. Answers are provided at end of chapter.

1. What is the fundamental difference between a survey and an experiment?

   ❑ **a.** surveys usually have larger sample sizes than experiments

   ❑ **b.** experiments have both treatment groups and a control

   ❑ **c.** surveys require random samples; experiments do not

   ❑ **d.** surveys are usually not analyzed statistically; experiments are always

2. In an experiment designed to evaluate the effects of a fertilizer on growth of a crop plant species in a variable outdoor environment, which of the following would be considered the dependent variable?

   ❑ **a.** average intensity of sunlight

   ❑ **b.** volume of irrigation water applied during the study

   ❑ **c.** yield

   ❑ **d.** mean temperature during the experiment

3. *True or F*alse. In an experiment designed to compare differences between means for a treatment group and a control, either the Student's *t* test or 95% confidence intervals are appropriate provided the assumption of normality is satisfied.

   ❑ **a.** True

   ❑ **b.** False

4. What is the null hypothesis in each of the two tests in Question 3?

   ❑ **a.** there is no difference in means of treatment vs control groups

   ❑ **b.** the two means are different, that is, they are from different populations

   ❑ **c.** there is a significant correlation between the two groups

   ❑ **d.** there is no correlation between the two groups

5. *True or False*. If the null hypothesis is rejected in a statistical test, one can conclude that the means of the two groups are significantly different.

   ❑ **a.** True

   ❑ **b.** False

6. If three or more means are to be compared, which of the following statistical tests is most appropriate?

   ❑ **a.** a Student's t test

   ❑ **b.** a correlation analysis

- ❑ **c.** an ANOVA
- ❑ **d.** a nonparametric chi-square test

7. *True or False*. If a correlation analysis indicates a significant correlation coefficient, it proves that one variable causes a measurable effect in the other variable.

- ❑ **a.** True
- ❑ **b.** False

8. *True or False*. If a regression analysis is significant at the 1% level, it is legitimate to predict values of the dependent variable outside the original range of the data.

- ❑ **a.** True
- ❑ **b.** False

9. If a Student's t test or ANOVA reports a P-value of <0.05, one can conclude all of the following EXCEPT ______________________________.

- ❑ **a.** the probability of a Type-I error is less than 5%
- ❑ **b.** at least one difference exists between the treatment means
- ❑ **c.** there is no statistical difference between the treatment means
- ❑ **d.** if the experiment were run repeatedly using the same methodology and sample size, one would expect to come to the same conclusions 95% of the time

10. If the assumptions of normality are not met in a set of data, they can be analyzed using one or more of the following EXCEPT

- ❑ **a.** nonlinear regression
- ❑ **b.** Kruskall-Wallace test
- ❑ **c.** sign test
- ❑ **d.** Wilcoxon rank sum test

Answers: 1) b; 2) c; 3) a; 4) a; 5) a; 6) c; 7) b; 8) b; 9) c; 10) a.

Reporting Research Results

Guidelines for Speaking and Writing Effectively

Chapter Learning Objectives

After studying this chapter, you should understand:

- The major means by which new scientific information is disseminated to the scientific community and general public.
- Important guidelines for reporting scientific information in oral presentations and written documents and publications.
- Some simple and effective means by which to evaluate and enhance your performance in public speaking and scientific writing.
- How to overcome some major obstacles to speaking and writing effectively.

Once a study or experiment has been conducted, analyzed and interpreted, the final critical step in completing a research project is to disseminate the new scientific information to other scientists and to the general public. This is normally accomplished by one or more of the following:

- **Press Releases**—preliminary reports of research findings that may be broadcast to the public via radio or television, or published in newspapers, magazines and similar printed media. Press releases are commonly compiled by persons other than the scientists who conducted the research and are rarely subjected to peer reviews, so it is important that the scientist(s) interviewed be allowed to review the article before it is broadcast or printed. If this is not possible, the scientist(s) involved should reconsider granting an interview.

- **Oral Presentations and Posters**—these are the two principal means for disseminating new information at scientific meetings and conferences, and are typically presented during timed sessions. Oral and poster presentations may involve either preliminary results of current research (in progress) or completed studies. In many instances, abstracts of oral and poster presentations are published in a *Conference Proceedings* for distribution to conference attendees.

- **Conference Proceedings**—these are published summaries of oral presentations and/or posters given at scientific meetings and conferences that are distributed to conference attendees, members of the sponsoring organizations and in many cases, to the general public. Conference proceedings may or may not be subjected to anonymous peer review prior to publication (see next section).

- **Unpublished Research Reports**—this category includes *laboratory reports* for undergraduate and graduate courses at academic institutions, and internal documents used by government agencies and private institutions to document research accomplishments involving highly sensitive or classified data which cannot be released to the public at the present time.

- **Theses and Dissertations**—reports of original research produced by graduate students at academic institutions. Most are eventually published as one or more articles in scientific journals.

- **Journal Research Articles**—journals are the principal means by which reports of original research (i.e., the *primary scientific literature*) are disseminated to the scientific community. Most research articles are published in refereed journals (in which submitted manuscripts are subjected to anonymous peer reviews) although a minority are published in nonrefereed journals. The anonymous peer review process has been used for many years to maintain quality in scientific publications and will be discussed in detail in Chapter 9.

- **Books, Book Chapters and Journal Review Articles**—these publications typically consolidate and summarize results of research that has been published previously and hence, are known as the *secondary scientific literature*. Journal review articles are typically subjected to the same types of anonymous peer review as research articles. Books and chapters of books dealing with scientific topics are commonly subjected to peer review, although the process may be different from that employed by refereed journals. Guidelines for publishing secondary scientific literature are discussed in Chapter 8.

Formats for Scientific Presentations and Publications

Reports of Original Research. Formats for presentations and publications of original research (i.e., the *primary scientific literature*) typically consist of a (an):

- Title Page or slide with list of authors and their affiliations,
- Abstract which summarizes the major points of the presentation or article (not generally included in oral presentations),
- Introduction which discusses the nature of the problem addressed, what is known about the problem (i.e., a literature review), gaps in knowledge (i.e., the basis for the current research), and objectives of the research reported on in the presentation or article,
- Materials and Methods section which discusses procedures used to accomplish each objective listed,
- Results section which summarizes the results of various experiments and statistical tests,
- Discussion section which discusses the author's interpretation of his or her research results and how they relate to previous studies and to the investigator's original hypotheses,
- Conclusions or Summary section which reviews the major points of the presentation or article,
- Acknowledgments section which acknowledges funding sources (if applicable) and any assistance provided by colleagues not included in author list, and
- Literature Cited section which includes all references cited in the text of the article (not generally included in oral presentations).

Techniques for preparing reports of original research are discussed in Chapter 7.

Reviews and Other Secondary Literature. In contrast to the format used in reports of original research (Chapter 7), review articles typically consist of the following sections:

- Title and author affiliations—similar to those for research articles,
- Abstract—similar to those for research articles except that discussion pertains to topics other than experimental results,
- Topic Headings and Subheadings—these summarize past and present research pertaining to each major and minor topic, and are placed in a logical sequence that will lead the reader to a conclusion or take-home message,
- Summary—optional

- Acknowledgment—included if applicable
- Literature Cited—includes all citations in text in the style preferred by the journal.

Techniques for preparing review articles and other examples of the secondary scientific literature are discussed in Chapter 8.

General Guidelines for Oral and Written Communications

The following general guidelines are applicable to all types of oral and written communications regardless of scientific content:

- **Oral presentations and written articles should be developed with a specific purpose in mind and directed to a specific audience.** For example, an oral presentation on disease transmission by mosquitoes presented to an audience of high school students would normally differ in both content and complexity from one presented to an audience of medical epidemiologists.

- **Presentations and documents should be well organized, easy to understand, and should provide the intended audience with a tangible take-home message.** Each section of a presentation or document should be written as clearly as possible, and each sentence or section should flow smoothly and logically into the next. Use of jargon and acronyms should be minimal—if used, they should be clearly explained at the time of first mention in the presentation or document. All figures or tables used in the presentation or document should be stand-alone, that is, self-explanatory and understandable by a typical member of the audience or reader without detailed explanation by the author(s). Collectively, the series of sections should provide the audience with a clear "take-home" message which may be stated explicitly in a *Summary* or *Conclusions* statement located near the end of the presentation or document.

- **Scientific communications of all types should be grammatically and semantically correct**. Most oral presentations today are developed using presentation software (e.g., Microsoft® Power Point) and are written using short phrases in a *bullet-type* format, whereas documents and manuscripts are written in complete sentences. In both cases, excessive wordiness should be avoided, and the author should ensure that word usage is correct, that all words are spelled correctly, and that sentences are constructed and punctuated properly. Be especially careful with words that are pronounced the same, for example, substitution of the word *insure* for *ensure* in the previous sentence would change its meaning entirely and probably make the author look amateurish. Many textbooks on grammar and style(s) are available (see References section in this chapter) and grammar/style checkers are included in word-processing packages (e.g., Microsoft® Word and others) and programs available from commercial vendors (e.g., http://www.Grammarly.com). Use of these products can improve the quality of written communications substantially.

- **Any materials included in the presentation(s) or document(s) that are owned or were developed by persons other than the author(s) should be acknowledged and documented.** Whether intentional or not, the inclusion of copyrighted material in a presentation or publication without proper documentation or permission of the owner(s) is unethical and may involve legal and/or civil penalties. Some simple strategies for avoiding unintentional *plagiarism* and other forms of academic misconduct are discussed in the chapter on academic integrity (Chapter 5).

Specific Guidelines for Scientific Presentations and Publications

All scientific presentations and articles should conform to the following specific guidelines:

- **All measurements should be reported in the metric system.**

- **Scientific names of biological organisms should conform to the *binomial system of nomenclature*, that is, *Genus* followed by the *species* epithet (and sometimes *subspecies*, if applicable).** Note that the genus name is always capitalized and the species epithet is always in lower case, and both are italicized. After first mention in the Abstract and in the text, the genus name is usually abbreviated but the species and subspecies continue to be spelled out in full. For example, at first mention in both the abstract and in the text, the scientific name of the boll weevil is written as *Anthonomus grandis* Boheman, but thereafter it is referred to as *A. grandis.* The last term of the name refers to the individual who first described the species.

- **Approved common names of biological organisms are written in lower case unless the name refers to a person or geographic location, and is never italicized.** For example, the insect referred to in the previous paragraph is correctly referred to as the boll weevil (rather than the Boll Weevil) while a distant relative associated with potatoes is known as the Colorado potato beetle, *Leptinotarsa decemlineata* (Say). The parentheses around the author name indicate that one or more nomenclatural changes in the scientific name have been made since the organism was first described.

- **Symbols and parameters in mathematical formulae must be clearly identified and explained. For example, the correct notation for a sample variance s^2 (see Chapter 3) is**

$$s^2 = \sum_{i=1}^{n} (x_i - \bar{x})^2 / N - 1$$

  where x_i = the value of individual observations, $\bar{x}$ = the sample mean, and N = total number of sample observations. This is an unbiased estimate of the population variance (σ^2). Note that all symbols and parameters have been identified and explained, and the reader does not have to refer back to Chapter 3 to understand the formula.

- **Conclusions should be based on statistical analyses of valid data, and statistics should be included in either the text or in figures or tables (but not both).** For example, results of a Student's *t*-test could be summarized in the text or in a figure as ($t = 16.5$; $df = 20$; $P < 0.001$) but not both because of redundancy.

- **Avoid overuse of the phrase "significant difference(s)"and never use the word "proves" when referring to results of statistical tests.** The term "significant difference" implies that the mean of one treatment group is consistently higher or lower than another group, and is expected to be evident a certain percentage of times in future tests provided that the same methodology is used in the tests (see Chapter 3). However, the detection of a "significant difference" between two or more means does not "prove" anything—it simply means that the test provided evidence that the null hypothesis of no differences among means should be rejected, i.e., real differences existed among two or more of the treatment means. Never use the term "proves" when referring to statistical analyses in scientific presentations or publications.

Developing, Evaluating and Enhancing Your Speaking Style

Development of oral presentations, like most other scientific communications, involves arranging slides containing data and other information on a particular topic in a logical order that conveys a clear take-home message to a specific target audience (this will be discussed extensively in Chapter 6). In order to be most effective, the speaker should

- **Be thoroughly familiar with the material in the presentation.** The only way to accomplish this is to read the background literature relating to the problem you are addressing and understand it thoroughly. When you deliver your presentation, you should strive to know more about the problem and the implications of your research than anyone in the room, including your professor.

- **Edit and rehearse your presentation to ensure that it conforms to time limits for your scheduled presentation.** Oral presentations at scientific meetings and conferences are typically given during timed sessions, and it is critical to complete your presentation within the allotted time slot. If you exceed your time period, you will irritate the moderator, the next scheduled speaker, and most of the audience, and you may have to terminate your presentation before you have made the points you had intended to make (which reflects poorly on you and your institution). The only way to avoid this type of embarrassment is to rehearse your presentation repeatedly before it is presented, and to ruthlessly edit it until it can be completed during the scheduled time slot. Ideally, you should give yourself a few additional minutes to answer questions from the audience.

- **Even if you are nervous, try to appear relaxed and confident**. As you approach the podium to deliver your presentation, walk at your normal pace and maintain good posture—avoid standing with one foot on top of the other while staring at the floor. Remember that this presentation and others to follow is what you worked so hard to be able to do, and this will be another important entry for your CV. This may come as a surprise to many, but unless you faint and fall off the podium, you are probably the only one who knows that you are nervous.

- **During the delivery, get off to a smooth start with a brief introductory statement that you have memorized.** A short opening statement of 2–3 carefully crafted sentences that you have memorized will usually serve to get your presentation going smoothly. Once you have begun your presentation, your slides should keep you "on-track" for the remainder of the presentation if they have been designed and organized correctly.

- **Speak clearly and loud enough to be heard by all members of the audience.** Before you begin to speak, ensure that your microphone is adjusted properly so everyone in the audience can hear you. Thereafter, speak in the manner you would use if you were explaining your work to a group of friends. Avoid speaking at an abnormally rapid rate for fear of running out of time—if you have rehearsed and edited your presentation carefully, this should not be a concern.

- **Vary your voice tones during the presentation.** There are few things that will put people to sleep faster than listening to a monotone voice in a dark room. Vary your voice tones while speaking and you will find that your audience will appear to be far more receptive to your presentation that they would have been otherwise.

- **Maintain eye contact with members of the audience and avoid reading excessively from notes or from the computer monitor.** As you are speaking, glance at your slides on the viewing screen or monitor, but avoid reading directly from slides or notes. Shift your eyes naturally to different members of the audience, but avoid staring at one person or group of people (this will make them nervous and make you look mechanical).

- **Avoid making irritating gestures (e.g., excessive movement of hands) while speaking**. Nervous gestures such as excessive movements of your arms, tugging on one of your ears or nose, and swaying back and forth while you are speaking is very distracting to the audience and detrimental to your presentation. Some simple techniques to identify and correct nervous gestures such as these are described in a later section of this chapter.

- **Be careful with laser pointers—never point toward audience.** Laser pointers generate concentrated visible light beams which can be very destructive to the eyes of humans and other animals. Never point laser pointers toward the audience. Also, avoid pointing an activated laser pointer at the walls or ceiling of the presentation room. If you do, you can be reasonably certain that a majority of your audience will be following the little red or green dot on the ceiling rather than watching your presentation. Hint: if your hands are "shakey," steady the laser pointer by holding it against a stationary point on the speaker's stand— this will eliminate any jerky movements of the pointer caused by nervousness.

- **Close your presentation with a summary or conclusions statement.** One of the best ways to end a presentation smoothly is to include a conclusions or summary slide near the end of the slide sequence. This will also serve to refresh the viewer's memories of the major points made in the presentation. The summary or conclusions statement may also be followed by an acknowledgments slide (required by many funding agencies) and a final slide soliciting questions from the audience.
- **Relax and try to interact well with members of the audience during the question portion of the presentation**. If time permits, answer questions that members of the audience have regarding your presentation. Be professional and avoid any hostile verbal exchanges with any person who appears to be "pushy" or confrontational. Indicate that you will be available for additional questions after the presentation session concludes.

Evaluating Your Performance. Regardless of their experience, all scientists should monitor their communication skills and make improvements whenever possible. Comments from colleagues attending a speaker's oral presentation(s) are helpful, but do not provide a means for the speaker to see himself or herself as others do. One effective way of accomplishing this is to use a simple technique that has been used by athletic teams to monitor their performance for years—record each of your rehearsals and your actual presentation with inexpensive video cameras that are currently components of most cell phones, tablet computers and many other electronic devices. Viewing a video of an oral presentation will allow a speaker to see himself as others do and to identify distracting or irritating mannerisms that he or she might not have otherwise been aware of (e.g., excessive movements of arms and hands, failure to maintain eye contact with members of the audience, and/or a dull monotone presentation style). By viewing and analyzing such videos, the speaker can modify his presentation style into one that he or she is comfortable with. Collectively, these procedures will enhance the oral communication skills of both inexperienced and experienced speakers and will go a long way in minimizing the speaking anxiety that plagues so many people.

Overcoming Obstacles to Speaking Effectively. One of the greatest impediments to effective communication among scientists is the fear of speaking in public. Known by various terms—*glossophobia*, *speaking anxiety disorder*, and many others—this unfortunate condition can seriously impede your ability to deliver effective oral presentations and posters, and may trigger more serious conditions such as panic attacks. Symptoms include dry mouth, sweating, increased heart rate, feelings of extreme anxiety or panic, quivering voice, and sometimes complete loss of voice. Because of the large number of people affected by this disorder, a large number of training courses and other remedial materials have been developed and are available from various sources (Table 4.1). Prospective scientists whose fear of speaking are severe and represent a true phobia would be well advised to take advantage of such remedial training opportunities or should consider an alternate career that does not involve public speaking.

A less severe but equally uncomfortable form of speaking anxiety is very common among inexperienced (and sometimes experienced) speakers who are unsure of themselves and/or have not mastered the material they are discussing in presentations. Several minutes before their presentation is scheduled to begin, they suddenly begin to feel wheezy, their arms and legs feel week, their mouths feel dry, and they feel like they are going to faint as they walk toward the podium to begin their presentation (many of these symptoms are similar to those of glossophobia, although not as severe). Telling speakers in this condition that such anxiety will usually disappear with experience may fall on deaf ears, but they might listen if you inform them that there are two very effective ways to prevent (or at least minimize) this type of speaking anxiety. As indicated previously, beginning the presentation with two or three memorized sentences at the time the title slide is projected will usually serve to start the presentation smoothly. Second, repetitive rehearsals during the week or so before (but not on the day of) the scheduled presentation will serve to reinforce the speaker's mastery of the subject matter

Table 4.1. Online sources of information on *glossophobia* – the fear of speaking in public.

Source	Title	URL
Dale Carnegie Training	Presentation Effectiveness	www.dalecarnegie.com
AllAboutCounseling	Glossophobia	www.allaboutcounseling.com
NativeRemedies	Glossophobia	www.nativeremedies.com
HxBenefit	Glossophobia	www.hxbenefit.com

of the presentation. If this is done routinely, the speaker will eventually begin to view himself as an expert (rather than a novice) and his audience as colleagues who are interested in his research (rather than ogres waiting for him to collapse on the podium).

Developing, Evaluating and Enhancing Your Writing Style

Manuscripts, reports and other printed documents are prepared in much the same way as oral presentations, that is, they are easily developed from outlines or templates, should be designed to provide a clear flow of logic leading to a tangible take-home message for the readers, and should conform to guidelines discussed in the previous sections of this chapter. However, the structure and hence, the preparation of written articles differs in two notable respects from that of oral presentations:

- oral presentations are typically presented during timed sessions and tend to be relatively brief; written articles are not constrained by time limitations and are typically longer, and
- written information in slides of oral presentations typically consists of short phrases in bullet statements; written articles consist of complete sentences, paragraphs and sections.

Thus, authors of written documents need to be particularly vigilant in terms of correct grammar, word usage, sentence structure and other elements of writing style. Some of the more common pitfalls in writing are discussed in the following examples and exercises:

Example 4.1—Use correct sentence structure

Incorrect: We grouped plants into 10 groups of five plants each and treated half of the groups (randomly selected) with fertilizer and deprived the other half of fertilizer, and all were irrigated at weekly intervals while being maintained under constant lighting conditions in the laboratory, and yields were measured after one month.

Correct: Ten groups of five randomly selected plants were established and maintained under constant lighting conditions (12L:12D) in the laboratory. Five randomly selected groups were subjected to fertilizer (the treatment group) and the remaining untreated groups were designated as the control. All plants were irrigated at weekly intervals, and yields within each group were measured after one month.

Note: The first sentence is extremely long and difficult to read. The corrected version consists of three sentences that are much easier to read and comprehend.

Example 4.2—Avoid Unnecessary Words

Incorrect: Once we analyzed the data, it was obvious that the addition of the fertilizer compound to the treatment groups resulted in significantly higher yields than the control groups which were deprived of fertilizer.

Correct: After a period of one month, the mean yield of the treatment group was significantly greater than the control ($t = 56.8$; $df = 48$; $P < 0.001$)

Note: Much of the wording in the first sentence is irrelevant and wordy. The corrected sentence is brief and to the point, and includes statistical results to support the conclusion.

Example 4.3—Use Correct Words or Terms.

Incorrect: Insects are the largest group of anthropoids …

Correct: Insects are the largest group of arthropods …

Note: In the first sentence, you are mistakenly informing the reader that insects are the largest group of the great apes (which includes humans), whereas the second sentence correctly identifies them as the largest group of the "jointed leg" animals which include the crustaceans, spiders, and many other taxa. One simple way to avoid embarrassing mistakes like this is to routinely use a dictionary when you are uncertain what a certain term should be.

Example 4.4—Identify acronyms at first mention in abstracts and text of written documents, and use them sparingly thereafter.

Incorrect: The USDA-ARS is responsible for …

Correct: The USDA Agricultural Research Service (ARS) is responsible for …

Note: Due to its size and importance, most U.S. citizens are probably familiar with the acronym of the U.S. Department of Agriculture (USDA), but this is not necessarily true for one of its agencies—the Agricultural Research Service (ARS), which has been identified correctly in the second sentence above. Citizens of many foreign countries may not recognize either acronym, so it may be wise to identify both in articles published in national or international journals, for example, … the Agricultural Research Service of the U.S. Department of Agriculture (USDA-ARS).

Example 4.5—Spell All Words or Terms Correctly.

Incorrect: We used an AONAV to compare the means for the treatments and control.

Correct: We used an ANOVA to compare the means for the treatments and control.

Note: Watch the spelling or you might be embarrassed when the paper is published.

Example 4.6—Use of a Grammar/Style Checker

The following excerpt from the first paragraph in Chapter 1 of this book was evaluated using a commercial grammar/style checker:

> "Although the terms *science* and *technology* are often used interchangeably, it is important to recognize an important distinction between the two. Science is derived from the Latin words *scientia* ("knowledge") and *scire* ("to know)" and refers to a systematic and rational approach to investigating natural phenomena through a combination of observations, experimentation and testing of tentative explanations or *hypotheses,* all of which are elements of the *scientific method* (Gower 2002).[1] Science is a philosophy that …"

The software returned a comprehensive report of spelling, grammar and style in the following format:

> Your average sentence length is somewhat long, which may make your writing difficult to follow. We advise you to read the tutorial on sentence length and structure.
>
> Your usage of technical words is above average for your educational level. However, you may still want to check out our tutorial on vocabulary development to improve your writing skills.

Helpful Technology

Most modern word processing software contains extensions for checking spelling and grammar in written articles. In addition, many of the software products that check for plagiarism also have provisions for assessing writing style and grammar (Table 4.2). Such products are effective and relatively inexpensive, and make writing less frustrating and a lot more enjoyable than it would be otherwise. Use of such products is highly recommended, particularly for those not proficient in spelling and rules of grammar and style.

Table 4.2. Grammar/Style Checker Software Available fromCommercial Vendors Online.

Grammarly	www.Grammarly.com
Style Writer Software	www.stylewriter-usa.com
English Grammar Checker Software	www.whitesmoke.com
Grammar Check Software	www.englishsoftware.com

Overcoming Obstacles to Writing Effectively

Few people have a morbid fear of writing (as do many of speaking in public) but many people dislike writing for a variety of reasons. *Writer's block* —the frustrating condition in which an author knows what he wants to say but can't seem to come up with the correct words for days or weeks—is one of the major impediments to effective writing in many people. Ways in which writer's block and other impediments can be avoided include the following:

- Allow yourself sufficient time to write an article—your mind tends to rebel and refuse to cooperate when you are tired or stressed,
- Use grammar/style checkers to enhance your writing skills,
- Gain proficiency in writing through practice—learn from reviews and comments you receive from colleagues,
- Use the outline and template method of writing (or similar technique), which will be presented and discussed in Chapters 7 and 8,
- Accept the fact that the more most people write, the more proficient they become and the more they enjoy writing.

Summary

1. Oral presentations and written articles should be developed with a specific purpose in mind and directed to a specific audience.

2. Presentations and documents should be well organized, easy to understand, and should provide the intended audience with a tangible take-home message.

3. Scientific communications of all types should be grammatically and semantically correct.

4. Any materials included in the presentation(s) or document(s) that are owned or were developed by persons other than the author(s) should be acknowledged and documented.

5. Evaluate and enhance your performance in speaking and writing as your career progresses.

References

Andersen, R. (1994). *Powerful writing skills*. New York: Fall River Press, 125 pp.

Hoffman, A. H. (2016). *Writing in the biological sciences: A comprehensive resource for scientific communication* (2nd ed.). New York: Oxford University Press, 329 pp.

Hoffman, A. H. (2010). Scientific writing and communication. New York: Oxford University Press, 682 pp.

Knisely, K. (2013). *A student handbook for writing in biology* (4th ed.). Sunderland, MA. Sinauer Associates, 318 pp.

Pechenik, J. A. (2013). *A short guide to writing about biology*. New York: Pearson, 276 pp.

Test Your Knowledge

Without referring to the material presented earlier in this chapter, mark an "x" in the box of the most appropriate answer to each question. Answers are provided at end of chapter.

1. Which of the following is an example of *primary scientific* lit*erature*?

❑ **a.** an unpublished laboratory report of an important research topic

❑ **b.** a proposal for research that has been designated as primarily important for an agency or institution

❑ **c.** journal article summarizing published research on a certain topic

❑ **d.** journal article reporting results of an original scientific experiment that has not been published previously.

2. *True or False*. The format and content for primary and secondary journal articles are essentially the same.

❑ **a.** True

❑ **b.** False

3. *Tru*e or False. In general, oral presentations and written articles should be directed towards a specific audience with similar interests.

❑ **a.** True

❑ **b.** False

4. Based on the rules of binomial nomenclature, which of the following represents the correct way of referring to a biological organism *at first mention* in the abstract and introduction of a scientific journal article?

❑ **a.** boll weevil, *Anthonomus grandis*

❑ **b.** boll weevil, *Anthonomus grandis* Boheman

❑ **c.** Boll Weevil, A. grandis Boheman

❑ **d.** A. grandis

5. Which of the following represents the correct way of referring to the same organism (see previous question) after it has been mentioned in the abstract and in the text of the article?

❑ **a.** boll weevil, *Anthonomus grandis*

❑ **b.** boll weevil, *Anthonomus grandis* Boheman

❑ **c.** Boll Weevil, A. grandis Boheman

❑ **d.** A. grandis

6. *True or False*. Results of statistical tests (Student's t or ANOVAs) should be summarized in the text, but never in tables or figures.

❑ **a.** True

❑ **b.** False

7. In general, what is the *null hypothesis* being tested when using a Student's *t* test or ANOVA?

❑ **a.** one or more treatment means differ significantly

❑ **b.** treatment means are from different statistical populations

❑ **c.** treatment means are from the same statistical population

❑ **d.** cannot be determined from this information alone

8. Which of the following is the most effective way to get an oral presentation off to a smooth start?

❑ **a.** recite two or three memorized sentences relating to your project

❑ **b.** tell several funny jokes

❑ **c.** introduce yourself and give a brief summary of your accomplishments and goals

❑ **d.** thank the sponsoring organization for the meeting and all of the colleagues who have assisted in your project

9. *True* or *False*. During the delivery of your oral presentation, you should read information from notes or from the projection screen as it is considered rude to look directly into the eyes of people in the audience.

❑ **a.** True

❑ **b.** False

10. Which of the following disorders is probably the single greatest impediment to effective public speaking?

❑ **a.** bipolar syndrome

❑ **b.** autism

❑ **c.** glossophobia

❑ **d.** delayed stress syndrome

Answers: 1) d; 2) b; 3) a; 4) b; 5) d; 6) b; 7) c; 8) a; 9) b; 10) c.

Importance of Scientific Integrity

An Imperative for All Scientists

Chapter Learning Objectives

After studying this chapter, you should understand:

- Why it is critical for all scientists to maintain high standards of academic integrity.
- Various forms of academic fraud including *data fabrication, data falsification,* and plagiarism, and why these are commonly referred to as "career killers."
- Simple techniques that you can use to protect yourself from unintentional plagiarism and other types of scientific misconduct that could quickly destroy your academic career.

Scientists specializing in various disciplines are involved indirectly or directly in virtually all aspects of modern society. Scientific research tends to be generative, that is, one important study commonly provides the basis for additional studies which continue until a particular goal is achieved. For example, the discovery of a new microorganism that appears to have antibiotic properties usually triggers additional studies (by the original group or others) designed to identify the antibiotic and evaluate its efficacy against a particular pathogen and its safety (for humans) as required for approval by the Food and Drug Administration (FDA). Such evaluations are nearly always based on the time-honored *scientific method* of experimentation and hypothesis testing which was discussed extensively in Chapter 4. If valid scientific research demonstrates that the antibiotic in question is indeed effective against the pathogen and is safe for humans, it is probable that a new treatment for the disease caused by the pathogen will be approved and become available to millions of people. On the other hand, fraudulent data at any level in the series of studies that led to approval by FDA could conceivably lead to the marketing of an ineffective or dangerous product that could be financially disastrous to the developer and detrimental to the health of millions of consumers. Because of the pervasive influence of science in modern society and the fact that scientific research and technology may have profound effects on human populations, scientists are not only expected to be competent, but also to have impeccable professional standards.

Fraud in Science

Although most scientists undoubtedly strive to maintain high professional standards, fraud and other forms of academic misconduct are probably more common than most of us would like to think (Dingell 1993). Several notorious recent examples of fraud in scientific research and their adverse consequences for the researcher(s) involved, the scientific community and public in general include the following:

- A respected American researcher on human obesity and the human aging process admitted in court that he had falsified data in 17 grant applications submitted to the National Institutes of Health (NIH) and had published at least 10 journal articles which contained fictitious data. He became the first American academic to be sentenced to prison for scientific fraud (*The New York Times*, October 25, 2006).[1]

- A fraudulent study conducted by a prominent British medical researcher reported a link between the measles, mumps and rubella (MMR) vaccination and the appearance of autism in humans. The British Medical Council declared the study to be fraudulent, the paper was retracted, and the researcher's medical license was revoked. This fraudulent research was instrumental in generating an anti-vaccination movement that continues to impede human disease control efforts in many areas of the world (*The New York Times*, January 12, 2011).[2]

- A fraudulent study conducted by a former Iowa State University researcher in which results of an HIV vaccination test were falsified resulted in the indictment and conviction of the researcher on two felony counts of making false statements on a government grant. The university he

---

[1] The EAIE Barometer. 2013. "Scientists sent to prison for fraudulent misconduct." Accessed June 15, 2015. http://www.universityworldnews.com/article.php?story

[2] WebMD. 2011. "BMJ declares Vaccine-Autism study 'an elaborate fraud'." Accessed March 8, 2016. http://www.webmd.com/brain/autism/news/

worked for was ordered to repay nearly $500,000 in grant money and an additional $1.4 million in remaining grant funds that had been approved for the project were cancelled by the National Institutes for Health (*The Federal Register*, December 23, 2013).[3]

Data Fabrication and Falsification. Each of the above cases were the result of either *data fabrication* (i.e., the deliberate creation and use of fictitious data) or *data falsification* (i.e., the deliberate manipulation or misrepresentation of data to influence the results of research reports). Data fabrication is commonly referred to as "dry-labbing" and essentially amounts to publishing the results of studies that were never conducted. Data falsification may include a variety of unethical activities including (1) the deliberate misrepresentation of procedures used in data collection and analysis when reporting results of research in oral presentations and/or publications, (2) the deletion of data points or observations from data sets without statistical justification in efforts to alter the results of field or laboratory experiments, and (3) any deliberate misrepresentations in oral and/or written communications that contribute to interpretations and conclusions that are knowingly false.

Plagiarism. Perhaps the most common form of academic fraud is *plagiarism*, which is defined as falsely claiming another person's ideas and/or writings as your own by failing to properly acknowledge the original source of the material in oral presentations and/or publications. Many cases of plagiarism are deliberate (e.g., knowingly claiming another scientist's ideas or writings as your own in scientific communications) while others may be unintentional (e.g., failing to properly cite a source because of forgetfulness or carelessness). Whether intentional or unintentional, plagiarism is considered to represent academic theft by virtually all scientists, and may have serious professional and/or legal consequences for those who engage in it.

Recipients of Federal grants who are convicted of academic fraud may have their names, institutions and the specifics of their cases published in the *Federal Register*, the *NIH* (National Institutes of Health) *Guide to Grants and Contracts*, the *PHS* (Public Health Service) *Administrative Actions Bulletin Board* and other media available to the scientific community and public. Price (2006) published a summary of academic misconduct cases handled by the U.S. Office of Research Integrity (ORI) between 1992 and 2005, which included the following three examples of plagiarism:

- An Associate Professor at the University of Southern California copied verbatim most of a proposal prepared by another scientist for submission to a state funding agency and included it in his own grant proposal which was submitted to NIH. Unfortunately for the Associate Professor, the original author of the proposal was a member of the NIH panel that reviewed the proposal, and recognized his own work. The plagiarist was sanctioned by the ORI for a period of three years[4].

- An Instructor in Medicine at Dana Farber Cancer Institute copied verbatim an NIH grant application that had been prepared by his former mentor at the Harvard Medical School and used it in his own grant application for HIV/AIDS research which was submitted to NIH. Unfortunately for the Instructor, a colleague of his mentor who had seen the original proposal was a member of the NIH review panel. The plagiarist was sanctioned by ORI for a period of three years.[4]

---

[3] The Des Moines Register. 2013. "AIDS vaccine test results faked." Accessed June 15, 2015. http://www.usatoday.com/story/news/nation/2013

[4] Price, A. R. 2006. Cases of plagiarism handled by the United States Office of Research Integrity 1992–2005. Pp. 46–56. In Plagiary: Cross-Disciplinary Studies in Plagiarism, Fabrication and Falsification. MI: MPublishing, University of Michigan Library, Ann Arbor, MI.

- An Instructor in neurobiology at Massachusetts General Hospital and Harvard Medical School copied an image of a Southern blot from a published manuscript (not his own) and presented it at a national society meeting. Unfortunately for the Instructor, the author of the manuscript which had been plagiarized was a member of the audience. The Instructor was sanctioned by ORI for a period of five years.[4]

Maintaining the Quality of Scientific Literature

Scientists who engage in data fabrication, data falsification and deliberate plagiarism appear to be oblivious to two aspects of modern science that provide considerable protection against the publication of shoddy and fraudulent research. First, virtually all major scientific journals use the time-honored system of anonymous peer reviews to ensure quality and soundness of all manuscripts published in their journal(s). Journal editors typically obtain 1–3 reviews from scientists who are recognized experts in their fields and whose identity is unknown to the author(s) of the submitted manuscript. Based on their reviews, the reviewers may recommend (1) acceptance of the manuscript in its present form, (2) acceptance with major or minor revisions, or (3) rejection of the manuscript for a variety of reasons including suspected or demonstrated plagiarism. The anonymous peer review process will be discussed extensively in Chapter 9.

A second protection against fraud results indirectly from two additional aspects of modern science: (1) valid scientific studies must be repeatable if they are indeed representative of "real world" trends and phenomena and (2) scientific studies tend to build on each other, that is, results from one study commonly provide the basis for subsequent studies on the same or related topics. If a scientist's research involves an important topic, it will almost certainly be repeated by other scientists in efforts to gain additional new information on that particular research topic. If an experiment is not repeatable by other scientists using similar methodology, the procedures and interpretations reported in the study will immediately become the focus of intense scrutiny by the scientific community, and the credibility of the researcher and his or her collaborators and institutions may be damaged or ruined if academic misconduct was involved.

Protecting Yourself Against Unintentional Plagiarism

Protecting yourself from deliberate academic fraud is simple—avoid it like the plague and make sure that all of your collaborators on grants and research projects do the same. If you have the misfortune of coauthoring a paper with another scientist who is engaged in fraudulent research during your collaborative research project, your credibility and reputation will very probably suffer the same fate as his or hers (this is the familiar situation of "guilt by association"). Protecting yourself from unintentional plagiarism is more problematical and requires an understanding of the various types of activities that constitute plagiarism, and the use of some simple techniques and inexpensive computer-age technology to help ensure that you do not become an unwitting plagiarist.

Reviewing and Summarizing Scientific Literature. When summarizing and comparing published literature on a given topic, record the bibliographic information for each source (author(s), date of publication, title, journal name, volume, and page numbers) at the time articles are read and critiques of each are written. Delaying this will increase the chances of inadvertently omitting one or more citations which should be included.

When taking notes, document all bibliographic information for each source at the time notes are taken, not afterwards,

- Avoid direct quotes in scientific articles,
- Avoid cutting and pasting of articles into your discussion,
- Use your own words to discuss aspects of each study or paper cited,
- Give yourself plenty of time and avoid working when tired.
- Routinely check your work with one of the many software packages designed to check for plagiarism and other factors.

Plagiarism Checkers. One of the most effective ways to protect yourself from unintentional plagiarism is to screen your written documents and manuscripts prior to submission with one of the many *plagiarism checkers* that are available from various vendors, for example, PaperRater® (www.paperrater.com), Turnitin® (www.turnitin.com), Grammarly® (www.grammarly.com), and many others. In addition to identifying possible plagiarism, many of these programs provide a variety of other functions helpful to writers (e.g., spelling, grammar, word choice and style), and most operate in a similar manner. In a typical example, a single paragraph from Chapter 1 of this text was uploaded into a designated window on the home page of a plagiarism checker and a report was requested (Figure 5.1). The software then compared the uploaded text with text in published journal articles, books, websites, and numerous other sources which were accessible via the Internet (one software vendor reports that its product has access to 10 billion or more online documents). The originality of text in this particular case was rated as high (83%) and the report indicated no evidence that it had been plagiarized from an existing document (Figure 5.1). When an abstract that had been copied verbatim from a published manuscript was uploaded into the same program, however, the report indicated a low degree of originality (0%) and included a warning of probable plagiarism with links to possible sources of plagiarized text (Figure 5.2). The latter provided access to the original manuscript which had been plagiarized and contained provisions for downloading the abstract and/or entire text (Figure 5.3).

Upload File Here

Science and Technology in Modern Society

Although the terms science and *technology* are commonly used in conjunction with each other, it is important to recognize an important distinction between the two. The term science is derived from the Latin words *scientia* ("knowledge") and *scire* ("to know)" and refers to a systematic and rational approach to investigating natural phenomena through a combination of observations, experimentation and testing of hypotheses (all of which are elements of the *scientific method* which will be discussed in greater detail in Chapter 3).

Science is thus a philosophy of inquiry that seeks to gain new knowledge of natural phenomena regardless of

PLAGIARISM CHECK

Originality: 83%

The text in this file appears to be original. No evidence of plagiarism is indicated.

Figure 5.1. Plagiarism check of selected paragraph from Chapter 1 of this text indicated a high level of originality (83%) and thus provided no evidence of plagiarism.

Upload File Here

ABSTRACT

Research was conducted to evaluate the effects of leaf excision and sample storage methods on spectral reflec-tance by foliage of giant reed, *Arundo donax*, an invasive weed which has caused extensive damage in many areas of the Rio Grande Basin in Texas and Mexico. Within 24 hours of excision, A. *donax* leaves exposed to ambient laboratory conditions (room temperature under natural lighting conditions) exhibited two trends indicative of physiological stress: 1) small but significant increases in reflectance of blue and red wavelengths (400–500 nm and 600–700

PLAGIARISM CHECK

Originality: 0%
PROBABLE PLAGIARISM – CHECK SOURCE(S) BELOW:
http://www.subplantsci.org/SPSJ/v63%202011/SPS63_07%20SUMMY%20ET%20AL%20-%20GALLEY%20FINAL_PDF.pdf

Figure 5.2 Plagiarism check of copied abstract indicated probable plagiarism (0% originality) and provided links to sources of original text.

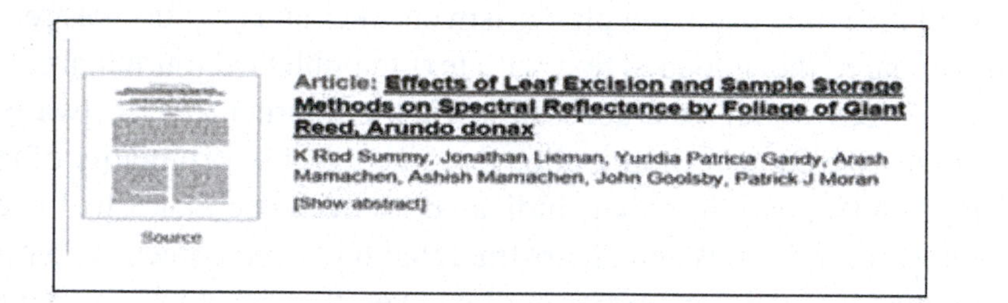

Subtropical Plant Science, 63:54-64.2011.

Effects of Leaf Excision and Sample Storage Methods on Spectral Reflectance by Foliage of Giant Reed, *Arundo donax*

K. Rod Summy,[1] Jonathan Lieman,[1] Yuridia Patricia Gandy,[1] Arash Mamachen[1] and Ashish Mamachen,[1] John Goolsby[2] and Patrick J. Moran[2]

[1] *The University of Texas – Pan American, Edinburg, TX 78539*
[2] *Agricultural Research Service, U.S. Department of Agriculture, Weslaco, TX 78596*

ABSTRACT

Research was conducted to evaluate the effects of leaf excision and sample storage methods on spectral reflectance by foliage of giant reed, *Arundo donax*, an invasive weed which has caused extensive damage in many areas of the Rio Grande Basin in Texas and Mexico. Within 24 hours of excision, *A. donax* leaves exposed to ambient laboratory conditions (room temperature under natural lighting conditions) exhibited two trends indicative of physiological stress: 1) small but significant increases in reflectance of blue and red wavelengths (400-500 nm and 600-700 nm, respectively) and 2) a substantial reduction in reflectance of near-infrared (NIR) wavelengths (700-1,100 nm). A similar but less pronounced trend was evident among leaf samples held within conventional paper sacks. Leaf samples held within sealed plastic bags (Glad-Bags) under two types of lighting conditions (natural light and artificial darkness) and temperature regimes (room temperature vs artificially cooled) exhibited slight but significant increases in both visible and NIR wavelengths (a trend that was also evident in attached foliage), although no evidence of physiological stress was detected during a 96-hour observation period. These trends indicate that accurate spectral measurements may be obtained from samples of *A. donax* foliage under for periods up to 72 - 96 hours following excision if such samples are transported and maintained in suitable containers designed to minimize effects of desiccation.

Additional Index Words: Carrizo cane, sampling, spectroradiometry, remote sensing.

Figure 5.3 Report provided a link that identified this publication as the source of the original text that had been plagiarized.

While plagiarism checkers provide an effective means to detect plagiarism in documents written by others (e.g., students in the classroom and collaborating scientists the laboratory), they are equally important in checking your own writings for unintentional plagiarism before sending them to their intended recipients. When writing lengthy reports or manuscripts, it is easy to inadvertently omit citations which should have been included in the document, but weren't. At best, such situations are embarrassing for the author(s) and may be considered by reviewers and journal editors as grounds for rejection of manuscripts submitted for publication. At worst, such omissions may be perceived by colleagues as deliberate plagiarism and damage the author(s) credibility as scientists. For these reasons, students and established scientists alike are well-advised to use one of the many plagiarism checkers that are available from various vendors routinely throughout their educational and scientific careers in order to protect their academic integrity.

Exercise 5.1

For each of the following situations, indicate whether or not the activity discussed constitutes plagiarism. If the activity constitutes plagiarism, suggest at least one remedy that would have prevented it from becoming an example of academic misconduct (answers are provided in the next paragraph).

a. In a paper discussing a new discovery involving DNA replication, the author mentions the *double-helix* structure of the DNA molecule without properly citing James Watson and Francis Crick's original paper on DNA structure in Nature. Does this constitute plagiarism? Why or why not?

____

____

____

b. An author of a paper involving climate change includes some imagery and documents downloaded from the website of the National Oceanic and Atmospheric Agency (NOAA) and does not cite the agency or authors of the documents (which were quoted verbatim) since NOAA is an agency of the U.S. government. Does this constitute plagiarism? Why or why not?

____

____

____

c. An author of a presentation on tropical diseases includes imagery of mosquitoes and other material downloaded from the Internet in a Power Point presentation, and does not cite the original source(s) since this is an oral presentation rather than a written document. Does this constitute plagiarism? Why or why not?

d. A professor publishes a paper based on one of his former student's Masters of Science thesis with himself as sole author because he has experienced great difficulty in locating the student after her graduation and wants to get the paper published. Does this constitute plagiarism? Why or why not?

e. A graduate student "cuts and pastes" a paragraph copied from a published manuscript into a file containing his draft PhD dissertation, but does not cite the original source since he changed the wording somewhat so it would not be a verbatim copy. Has he plagiarized the original document? Why or why not?

f. A scientist submits manuscripts based on the same data but with minor differences in wording to several journals on the grounds that the papers will reach a wider audience than they would have otherwise. Does this constitute plagiarism? Why or why not?

Summary

1. Scientific research plays a critical role in modern society, and it is imperative that all scientists maintain impeccable standards of academic integrity in reporting results of their research.

2. Two forms of academic misconduct—data fabrication and data falsification—are deliberate acts that may result in criminal penalties for fraud.

3. Deliberate plagiarism is probably the most common form of academic misconduct by students and scientists, and may result in civil penalties for copyright infringement and other serious charges.

4. The three forms of academic conduct—data fabrication, data falsification and plagiarism—are commonly referred to as "career killers" for obvious reasons.

5. Software known as plagiarism checkers should be used by all writers to check their work and minimize the possibility of committing unintentional plagiarism.

References

Dingell, J. D. (1993). Misconduct in medical research. *The New England Journal of Medicine,* 328, 1610–1615.

Judson, H. F. (2004). *The great betrayal: Fraud in science*. Orlando, FL: Harcourt, Inc., 480 pp.

Price, A. R. (2006). Cases of plagiarism handled by the United States Office of Research Integrity 1992–2005. *In Plagiary: Cross-disciplinary studies in plagiarism, fabrication and falsification*. (pp. 46–56). Ann Arbor, MI: MPublishing, University of Michigan Library.

The EAIE Barometer. (2013). *"Scientists sent to prison for fraudulent misconduct."* Retrieved June 15, 2015, from http://www.universityworldnews.com/article.php?story

WebMD. (2011). *"BMJ declares Vaccine-Autism study 'an elaborate fraud'."* Retrieved March 8, 2016, from http://www.webmd.com/brain/autism/news/

The Des Moines Register. (2013). *"AIDS vaccine test results faked."* Retrieved June 15, 2015, from http://www.usatoday.com/story/news/nation/2013

Test Your Knowledge

Without referring to the material presented earlier in this chapter, mark an "x" in the box of the most appropriate answer to each question. Answers are provided at end of chapter.

1. *True or False.* Because of the importance of science and technology in modern society, academic misconduct by scientists is a rare occurrence.
   - ❑ **a.** True
   - ❑ **b.** False

2. All of the following are forms of *academic misconduct* EXCEPT ________________.
   - ❑ **a.** reporting research results from a poorly designed experiment
   - ❑ **b.** reporting results of research that was never conducted
   - ❑ **c.** reporting data collected by another scientist without his or her knowledge
   - ❑ **d.** "tweaking" data to make results "look better" than they really are

3. Which of the following best describes the term *data fabrication*? ________________.
   - ❑ **a.** reporting research results from a poorly designed experiment (e.g., no control)
   - ❑ **b.** reporting results of research that was never conducted
   - ❑ **c.** reporting data collected by another scientist without his or her knowledge
   - ❑ **d.** "tweaking" data to make results "look better" than they really are

4. Which of the following represents an example of *data falsification*?
   - ❑ **a.** reporting research results from a poorly designed experiment (e.g., no control)
   - ❑ **b.** reporting results of research that was never conducted
   - ❑ **c.** reporting data collected by another scientist without his or her knowledge
   - ❑ **d.** "tweaking" data to make results "look better" than they really are

5. Which of the following is a blatant example of *deliberate plagiarism*?
   - ❑ **a.** reporting research results from a poorly designed experiment (e.g., no control)
   - ❑ **b.** reporting results of research that was never conducted
   - ❑ **c.** reporting data collected by another scientist without his or her knowledge
   - ❑ **d.** "tweaking" data to make results "look better" than they really are

6. *True or False.* Because of the sheer volume of scientific publications, plagiarism and other forms of academic misconduct are extremely difficult to detect.
   - ❑ **a.** True
   - ❑ **b.** False

7. *True or False*. Plagiarism is not generally detrimental to one's career as long as it is unintentional.
   - ❑ **a.** True
   - ❑ **b.** False

8. Which of the following is probably the best way to protect yourself against unintentional plagiarism?
   - ❑ **a.** carefully edit and reword text "cut and pasted" from published research
   - ❑ **b.** proofread your work carefully to make sure you properly cited authors of work cited in your text
   - ❑ **c.** routinely screen your writings with programs like PaperRater® and Turnitin®
   - ❑ **d.** never write when you are tired

9. *True or False*. Plagiarism applies primarily to written material (in which there is a paper trail) and is generally of no concern in oral presentations that are not videotaped
   - ❑ **a.** True
   - ❑ **b.** False

10. *True or False*. One of the major gaps in enforcing integrity in scientific research relates to the current lack of any real legal penalties for violators.
    - ❑ **a.** True
    - ❑ **b.** False

Answers: 1) b; 2) a; 3) b; 4) d; 5) c; 6) b; 7) b; 8) c; 9) b; 10) b.

Scientific Presentations and Publications

Oral Presentations and Posters

Communications at Scientific Meetings and Conferences

Chapter Learning Objectives

After studying the materials in this chapter, you should understand:

- The role and importance of oral presentations and posters in the dissemination of scientific information at meetings and conferences.
- How to develop effective oral and poster presentations, and to deliver them effectively to their intended audiences.
- Some simple techniques that will allow you to evaluate and enhance your speaking and writing skills.

For hundreds of years, scientific societies representing most disciplines have sponsored periodic meetings and conferences to discuss issues of importance to their memberships and to disseminate new knowledge to the scientific community. Aside from informal meetings and discussions among scientists, the formal exchange of ideas and new information at scientific meetings and conferences is accomplished primarily by two means—oral presentations and posters—both of which involve a combination of writing and speaking.

In preparation of their annual meetings and conferences, modern scientific societies usually announce a "call for abstracts" of oral presentations and posters that will be delivered at the meeting or conference (Figure 6.1). Submitted abstracts are generally reviewed by a screening committee to ensure that the subject matter is appropriate for the meeting and that the abstract conforms to the society's presentation policies. If approved, abstracts are commonly printed in the meeting program and in many cases, manuscripts of presentations and posters may be published in a *Conference Proceedings* which is subsequently distributed to the society's membership and thus becomes available to the scientific community in general.

The remarkable advances in computer technology which have occurred during the past several decades have dramatically changed the ways in which oral presentations and posters are developed and delivered. In contrast to the time in which the state-of-the-art technology involved the use of "stick-on

Call for Abstracts

The Society of Hypothetical Science will hold its 69th Annual Meeting in Somewhere, TX on 6–9 February 20XX. Researchers, students and other interested parties wishing to present oral presentations or posters should submit an abstract for approval to rksuper@shs.edu by January 31, 20XX. Approved abstracts of presentations and posters will be published in the meeting program and will be posted on the society's website www.xhs.edu.

Students who submit posters AND present a 3-minute (oral) communication to share their results will be eligible to receive one of the awards for the top 3 places (abstracts of posters that are not presented will be published but will not be eligible for an award).

Abstract format: Title (bold) Author(s) (bold) *Affiliations (italics; not bold; one affiliation per line;* include corresponding author's email address*)* Abstract (normal font). Limit to 1 paragraph of no more than 275 words.

Poster format: Dimensions: The poster should NOT exceed 36"Wx48"L (91x122 cm).Title lines: Brief but informative title, authors' names and affiliations. Use a simple sans serif-face font such as Arial. Lettering for the title should be at least 1 inch; the authors' names and affiliations may be somewhat smaller. Content: Please use tables and figures and limit verbage; research details can be provided in discussions with interested parties; lettering for text and illustrations should be legible from a short distance. Display: A poster board (~36"Wx48"L) will be provided for mounting each poster. For further information about the poster session, contact: rnsuper@shs.edu and indicate INQUIRY in the subject line.

Figure 6.1 Typical call for abstracts issued by a scientific society prior to an Annual Meeting.

letters and numbers, overhead projectors with acetate data sheets and/or slide projectors loaded with 35mm color-positive slides, today's sophisticated presentation software (e.g., Microsoft Power Point®) provides the means to develop powerful oral and poster presentations in a fraction of the time that was required just a few years ago. The thought processes and communication skills that go into the development of oral and poster presentations, however, remain the same as they have always been and require considerable planning and organization on the part of the presenter(s). Specific guidelines for the development and delivery of oral and poster are discussed in the following.

Oral Presentations

As with all types of scientific communication, the obvious first step in the development of oral presentations is to select a topic of importance that will be presented to a specific target audience. Collectively, these two factors will largely determine both the content and level of complexity of the presentation. For example, if the topic involves "the incidence of dengue fever (a mosquito-borne tropical disease vectored by mosquitoes) in southern United States and northern Mexico," the content and level of complexity of a presentation delivered to medical specialists will differ substantially from one designed for politicians and/or city planners, and both will differ from a presentation designed for students in a middle-school science class. Two other factors that are especially important considerations for oral presentations are the nature of the event at which the presentation will be delivered and the time slot that will be available to the speaker at the time of presentation. Collectively, these two factors will largely determine the number of slides to include in the presentation. The length of a presentation (determined by the number of slides included and the time allocated for each) may be irrelevant if a single speaker has a one-hour (or unlimited) time slot, but becomes immensely important in 15-minute presentations at scientific meetings with concurrent sessions. Moderators at such sessions are responsible for maintaining presentations on a rigid schedule, and it is imperative for speakers to deliver their presentations within the allotted time slot. A good rule-of-thumb is to allow approximately one minute per slide with an additional 3–5 minutes at the end to answer questions by the audience.

Organizing an Oral Presentation. Once the number of slides to include in a presentation has been determined, the next step is to develop an outline that describes the logic of the presentation and content of each slide (Figure 6.2). Virtually all oral presentations begin with a title slide that includes the title of the presentation followed by the name(s) of the author(s) and their affiliation(s) (Figure 6.3a). The

Outline for Presentation

Slide	Description
1	Title slide
2	Important points
3	Inclusion of Figures
4	Keep Slides Simple
5	Summary Slide
6	Thank you – Questions?

Figure 6.2 Outline of oral presentation indicating order and content of slides.

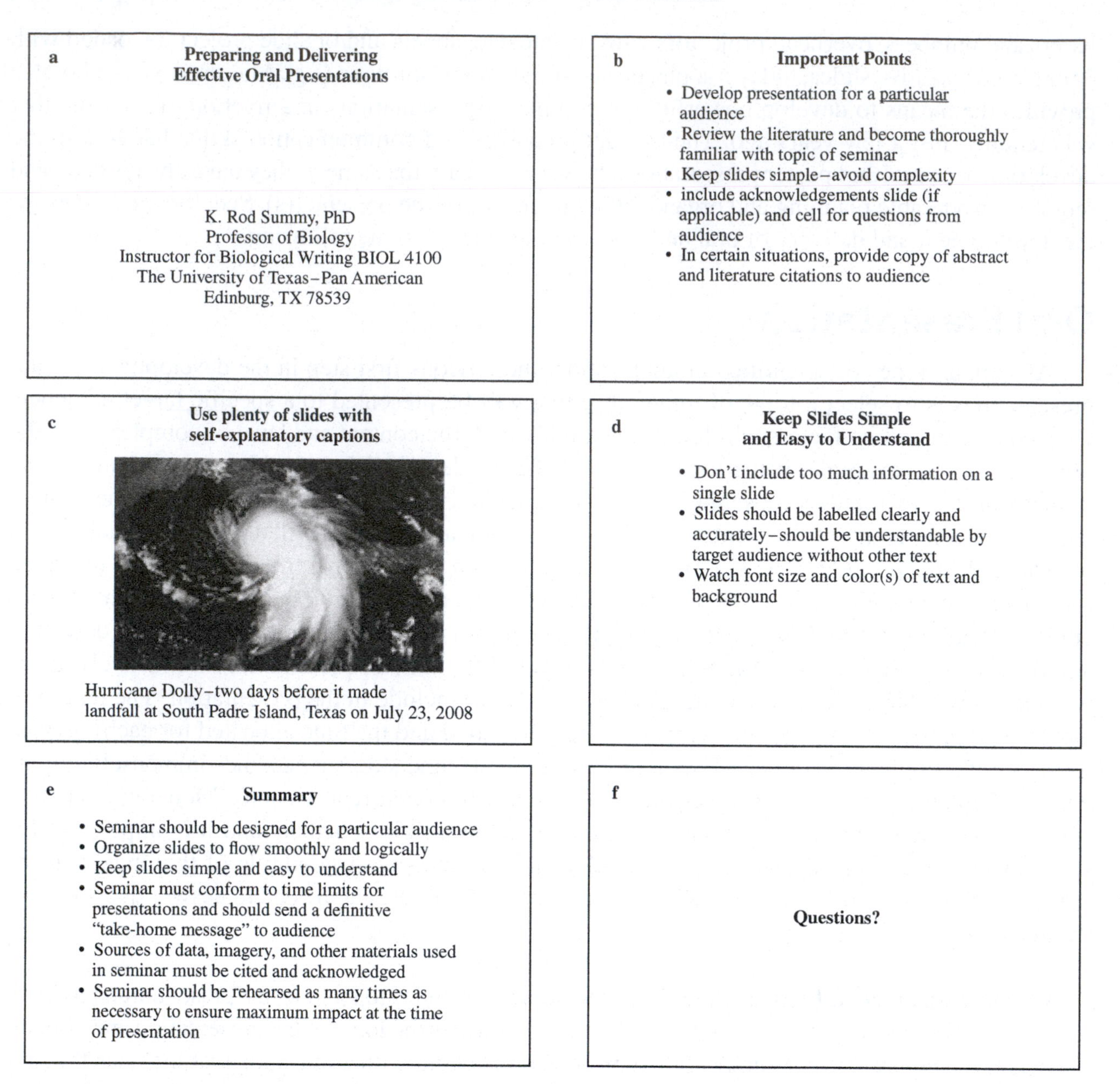

Figure 6.3 Portion of a typical oral presentation using Microsoft® Power Point software. Presentations begin with a title slide and should end smoothly with a summary and final slide. Slide sequence usually includes title slide (a), topic slides (b–d), summary slide (e), and terminal slide that ends presentation smoothly and opens discussion with audience (f).

remaining slides should focus on various aspects of the topic under discussion in a logical manner that leads to a clear "take-home" message for the audience (Figure 6.3b–d). The take-home message may be stated explicitly in a "Summary" slide that briefly reviews the information presented and states the author(s) conclusion(s) and/or recommendation(s) (Figure 6.3e).

Presentations with similar content but different levels of complexity may convey a variety of take-home messages—for example, a presentation on dengue fever delivered to an audience of medical specialists might recommend a need for additional research on vector transmission of the virus, while a similar presentation delivered to politicians and/or city managers might recommend the development of contingency plans for mosquito abatement in areas prone to flooding caused by tropical storms, while another delivered to middle-school students might promote medical entomology as a career option. Summary slides are commonly followed by an "Acknowledgments" slide which expresses appreciation to colleagues who provided substantial assistance during the course of a study and/or support provided by various funding agencies (many funding agencies provide a "boilerplate" statement for this purpose). The final slide in most oral presentations is a slide with the phrase "Thank you—Questions?" or something similar (Figure 6.3f) which serves to end the presentation smoothly and opens the discussion for questions from the audience.

The following guidelines apply to all slides included in an oral presentation:

- Slides should be simple, uncluttered, and contain only information relevant to the topic of the presentation,

- Slides should be self-explanatory or "stand-alone," that is, a knowledgeable viewer should be able to understand the information contained in the slide without requiring a detailed explanation from the presenter,

- Slides should be pleasing to view, and the text and background colors should contrast pleasingly, for example, yellow or white text on a blue or green background would normally be a pleasing combination to most audiences,

- Certain color combinations should be avoided, for example, red text on a purple background purple would strain the eyes (and patience) of most viewers, whereas red text on a green background (or vice versa) would probably not be distinguishable by members of the target audience with red-green color blindness,

- Slides should be arranged in a logical order such that the information summarized in one slide flows smoothly and logically into the next, for example, a slide listing several gaps in knowledge on a certain topic may serve as a prelude for objectives listed in the following slide,

- Any data or imagery contained in slides that are not the property of the preparer(s) or presenter(s) must be acknowledged by an annotation located somewhere within the slide or in the title caption.

Slides containing tables and/or figures with statistical data should conform to the following additional guidelines (Figure 6.4 and 6.5).

- Titles or captions for tables are normally located above the main body of the table, whereas captions for figures are normally located below the figure,

- Titles or captions for tables and figures should be self-explanatory and all parameters included in tables and figures must be labelled and explained by means of footnotes or explanations provided in the table or figure captions, that is, the slide must be "stand-alone."

- Statistical data presented in tables and graphs should include not only the sample estimates, but also measures of variability (e.g., standard deviations, standard errors, and/or confidence intervals or error bars),

Original Table	Problems

Original Table

Table 1. Data from experiment 1 (2010) which showed that treatment 1 produced the highest yields.

Treatment	Average Yield
1	24.8
2	12.2
3	14.9
C	8.5

Problems

- Table title provides essentially no information regarding the nature of the comparisons summarized in the table.
- Column lables provide no information regarding identity of variables listed.
- Information regarding sample size and units of measurement lacking.
- No measures variability of data included.
- Results of statistical omitted.

Revised Table

Table 1. Comparison of yields for cucumbers (var. "Burley") subjected to three levels of nitrogen with untreated control (2010).

Trt	N	Mean yield[1] (kg/ha)	SD
1	25	24.8a	4.2
2	20	18.2b	2.1
3	20	15.3b	5.5
C	25	10.5c	1.2

[1] Means of followed by same letter not significantly different as 5% probability level (Tukey's pairwise comparison test).

Improvements

- Title descriptive—provides adequate information regarding nature of experiment and identity of variables.
- Units of measurement clearly indicated in metric system (kg\ha).
- Measure of variability included (standard deviation).
- Results of statistical analysis included (analysis of variance with Tukey's pairwise comparison test)—footnote with standard boilerplate statement.

Figure 6.4 Comparison of poorly designed (upper) and acceptable (lower) "stand-alone" tables for reporting statistics in oral and poster presentations.

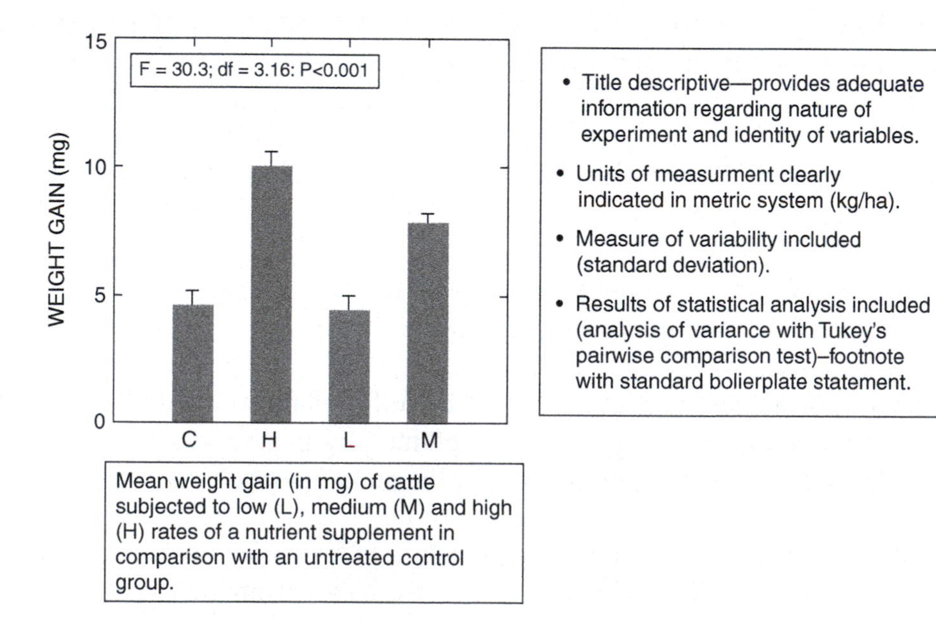

Figure 6.5 Example of an acceptable stand-alone bar graph for statistical comparisons in oral and written presentations.

- Statistical tests used in data analyses must be identified and results of tests should be explained by means of footnotes or explanations provided in table or figure captions. Symbols used to indicate differences among sample estimates must be explained and the probability level identified in footnotes or explanations provided in captions, for example, a statement such as ". . . means followed by same letter are not significantly different at 5% probability level—Tukey's pairwise comparison test,"

- Avoid including too many comparisons or trends in line graphs—four or five lines is generally sufficient, and the inclusion of more usually causes confusion and difficulties in interpretations.

Rehearsing and Editing. The importance of rehearsing and editing an oral presentation before its actual delivery at a scientific meeting cannot be overemphasized. This is the only reliable means of determining how long a presentation will actually take when it is delivered to its intended audience, and is also the best way to train yourself to deliver the presentation smoothly. Time yourself during each rehearsal and edit ruthlessly (if necessary) after each rehearsal—delete any slides that are unnecessary or detract from the presentation, for example, "crowded" slides containing too much information. The author would recommend at least three or four rehearsals during the week or so before the presentation, but no rehearsals (and especially no changes) on the day the presentation will be delivered. At that time, you should be thoroughly familiar with your presentation and the ideas you would like to convey to your audience, and any last-minute changes (unless they are absolutely necessary) will probably throw you "off track" and detract from the presentation.

Delivering the Presentation. Oral presentations should begin smoothly when the title slide is projected and end gracefully with the final slide. One practice that has always served the author of this book well is to memorize several opening statements that will be made while the title slide is displayed—this usually allows the speaker to explain the nature of the topic and the objectives of the presentation without fumbling for words. Once the presentation has begun, the information contained in the slides of a well-designed presentation will usually enable the speaker to stay "on track" and make the points that the presentation was designed to make. A very effective way to end the presentation smoothly is to include a summary slide near the end followed by an acknowledgment slide (if applicable) and a final slide that effectively ends the presentation and opens the floor for questions from the audience (Figure 6.3e,f).

The following suggestions were discussed earlier (Chapter 4), but are worth restating. During the actual delivery of your presentation:

- **Breath deeply and relax**—there is no need to rush if you have rehearsed your presentation sufficiently and are confident that it can be delivered within the designated time slot,

- **Speak clearly and loud enough** so that all members of the audience can hear and understand what you are saying,

- **Avoid excessive (and possibly irritating) hand and body motions** while speaking,

- **Maintain eye contact** with members of the audience as much as possible,

- **Read information from slides projected on the screen**—avoid reading excessively from note cards or the computer monitor,

- **If you are using a laser pointer, avoid pointing it toward the audience**—laser pointers generate concentrated light energy and can damage the retina of human eyes,

- **When using a laser pointer, stabilize the device** by resting the base on a stable surface and activate it only when pointing to specific points on the screen—never allow a laser beam "roam" across the screen or onto the walls and ceiling of the presentation room—this is very distracting to the audience,

- If no laser pointer is available, the cursor on the computer monitor can be used to point to data and other information on slides.

Some simple techniques to evaluate and enhance your performance in public speaking are discussed in Chapter 4. In addition, professional sources of information dealing with fear of speaking (*glossophobia*) are provided, although knowing your material thoroughly and rehearsing repeatedly prior to the delivery will go a long way in addressing this problem.

Poster Presentations

Poster presentations represent the second major type of formal presentations at scientific meetings and conferences. Posters are typically developed using word-processors (e.g., Microsoft® Word or equivalent) or publishing software (e.g., Microsoft® Publisher or equivalent) and printed on 2x3 or 3x4 ft sheets of presentation quality poster paper using large-format printers. Because of space limitations, the text portion of the poster is typically brief and figure captions (which are usually printed in smaller font) may be used to explain details of topics mentioned in the text (this technique usually increases the potential information content of the poster considerably). Most or all of the guidelines discussed previously for oral presentations are also applicable to poster presentations, although posters are subject to a few additional considerations:

- Posters should be prepared according to instructions provided in the "call for abstracts or paper" issued by the sponsor of the meeting or conference,

- The title of the presentation is most commonly centered at the top of the poster in relatively large font, with the author(s) and their affiliations listed below in smaller font,

- Posters discussing complete or preliminary studies may be written in typical manuscript format, that is, with an abstract followed by an introduction, materials and methods, results and discussion, acknowledgments (if applicable) and references cited sections,

- Posters discussing topics other than research (e.g., concepts, programs, ecological problems, etc.) may be written in other styles permitted by the sponsor of the meeting,

- Table and figure captions may be used to expand the information content of the poster substantially, for example, instead of providing a description of an instrument in the text of the methodology section, reference a figure which provides the same information in smaller font size,

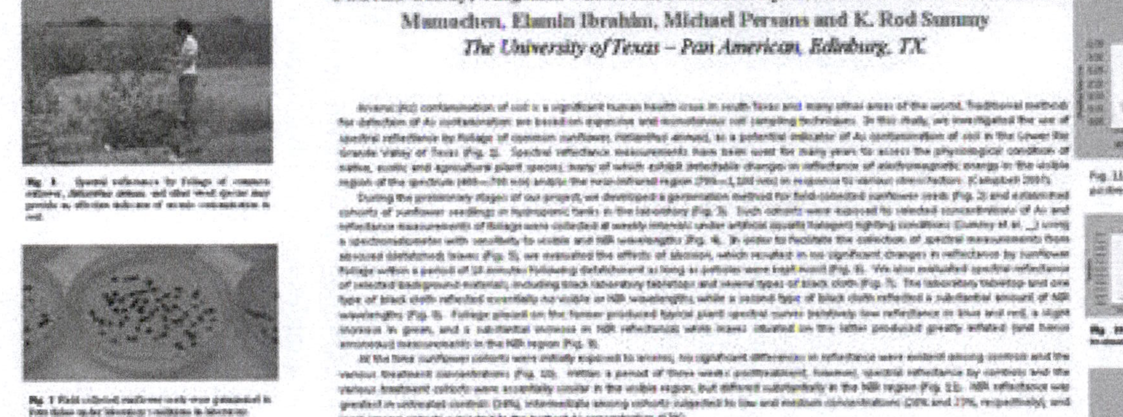

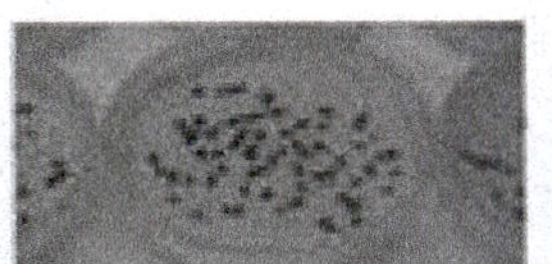

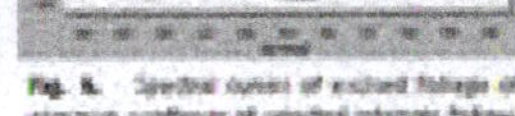
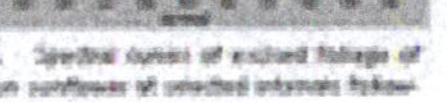
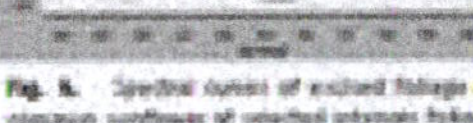

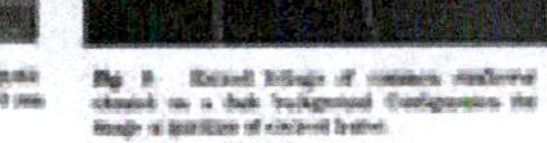

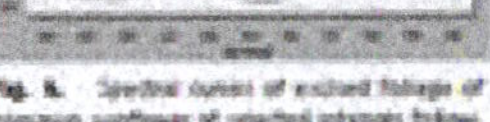
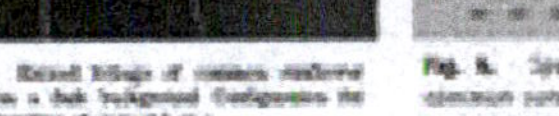

Figure 6.6 Example of a well-balanced poster presented at a scientific conference.

- Figures and tables should be positioned within the text so that the poster will have a pleasing and "balanced" appearance (Figure 6.6),

- Posters should be printed on quality presentation poster paper and transported to meetings in poster tubes to prevent damage during transit,

- Posters are generally presented to small groups of individuals during designated poster sessions at meetings and conferences, and thus are not as "time critical" as oral presentations—content of posters is far more important.

During the scheduled poster session, it is imperative that at least one of the authors of the poster be present to discuss the poster contents with interested viewers. Representatives of academic and research institutions frequently attend poster presentations to meet and talk with potential recruits for their programs and institutions. Be at your best when speaking with viewers who appear to be interested in your work as they may turn out to be your future employers.

Exercise 6.1

Develop a 10-minute oral presentation involving a topic of your choice and make a video of yourself delivering the presentation alone or with a group of friends. Make notes of any mannerisms that you would like to enhance or eliminate while speaking, and make a concerted effort to avoid those mannerisms during a second or third presentation that is recorded on video. You will be surprised how effective this method is in projecting yourself as you would like others to perceive you when speaking.

Exercise 6.2

Develop simple poster on a topic of your choice using Microsoft® Power Point and print it on an 8 ½ x 11- inch sheet of paper. Use the text to make general statements about your topic and the captions for figures and tables to provide explanations of instruments, methodologies and results of experiments. Arrange the figures and tables to give a balanced and pleasing appearance to the poster.

Suggested Resources

Andersen, R. (1994). *Powerful writing skills.* New York: Fall River Press, 125 pp.

Hoffman, A. H. (2016). *Writing in the biological sciences: A comprehensive resource for scientific communication* (2nd ed.). New York: Oxford University Press, 329 pp.

Hoffman, A. H. (2010). *Scientific writing and communication.* New York: Oxford University Press, 682 pp.

Knisely, K. (2013). *A student handbook for writing in biology* (4th ed.). Sunderland, MA: Sinauer Associates, 318 pp.

Pechenik, J. A. (2013). *A short guide to writing about biology.* New York: Pearson, 276 pp.

Test Your Knowledge

Without referring to the material presented earlier in this chapter, mark an "x" in the box of the most appropriate answer to each question. Answers are provided at end of chapter.

1. *True or False*. In general, the most effective oral presentations are those designed for a wide audience rather than a restricted group of specialists.

   ❑ **a.** True

   ❑ **b.** False

2. How many slides would be most appropriate for a 15-minute oral presentation at a scientific meeting with concurrent sessions?

   ❑ **a.** 5

   ❑ **b.** 10

   ❑ **c.** 15

   ❑ **d.** 20

3. Which of the following will usually serve to get an oral presentation off to a smooth start?

   ❑ **a.** read the title slide to the audience

   ❑ **b.** memorize a 2–3-sentence opening statement

   ❑ **c.** tell a joke or funny story

   ❑ **d.** let things fall into place on their own

4. *True or False*. One of the most effective ways to discuss data during an oral presentation is to read data directly off the monitor or projection screen.

   ❑ **a.** True

   ❑ **b.** False

5. *True or False*. Since time is limited in most oral presentations, it is usually best to include as much data as possible on slides.

   ❑ **a.** True

   ❑ **b.** False

6. A slide that is *stand-alone* is one that is ________________________.

   ❑ **a.** self-explanatory

   ❑ **b.** displayed separately from the remainder of the slides in the presentation

   ❑ **c.** highly complex and requires a detailed explanation by the presenter

   ❑ **d.** copyrighted

7. Why would it be unwise to use a color combination of red letters on a green background for data slides in a presentation?

❑ **a.** this combination is known to make many viewers nauseous

❑ **b.** it would probably be difficult to see under low lighting intensities

❑ **c.** slide would not be discernible by viewers with color *blindness*

❑ **d.** many societies discourage or prohibit nontraditional colors at presentations

8. *True or False*. Poster presentations are generally discouraged at scientific meetings because they are considered to be less impactive than oral presentations.

❑ **a.** True

❑ **b** False

9. *True or False*. During poster sessions at scientific meetings, it is not usually necessary for the author(s) to be present as posters are displayed in a manner that is readily accessible to viewers.

❑ **a.** True

❑ **b.** False

10. *True or False*. Due to the limited space available on most posters, the use of images and data figures is discouraged.

❑ **a.** True

❑ **b.** False

Answers: 1) b; 2) b; 3) b; 4) b; 5) b; 6) a; 7) c; 8) b; 9) a; 10) b.

Reports of Original Research

The Primary Scientific Literature

Chapter Learning Objectives

After studying this chapter, you should understand:

- The various categories of publications that constitute the *primary scientific literature*.
- The basic structure and format of journal articles and other publications reporting results of original research.
- How to write each section of journal articles and other research reports efficiently and effectively and avoid "writer's block."

The *primary scientific literature* consists of articles from a variety of sources that report research findings that have not been published previously. The primary scientific literature is commonly grouped into several broad categories:

- **Theses and Dissertations**—reports of original research conducted by graduate students to fulfill requirements for Master of Science degrees (theses) and Doctor of Philosophy degrees (dissertations). Most theses and dissertations are eventually published as one or more articles in scientific journals.

- **Unpublished Research Reports**—this category includes *laboratory reports* prepared by students for undergraduate and graduate courses in academic institutions, and internal documents used by government agencies and private institutions to document research accomplishments that have not been officially released to the public. Access to the latter may require security clearances or other permissions.

- **Conference Proceedings**—these are published summaries of oral presentations and/or posters given at scientific meetings and conferences that are distributed to conference attendees, members of the sponsoring organizations and in many cases, to the general public. Conference proceedings may or may not be subjected to anonymous peer review prior to publication (see next section).

- **Journal Research Articles**—journals are the principal means by which reports of original research are disseminated to the scientific community. Articles submitted to most major research journals are subjected to rigorous peer reviews by anonymous experts selected by the editor, in which case the articles are considered to be *refereed* articles. Articles published in journals that do not use the anonymous review system are considered to be *nonrefereed* and are generally considered to be less impactive than refereed articles. The peer review process has been used for centuries as a means for maintaining quality in scientific publications and will be discussed extensively in Chapter 9.

Structure and Format of Journal Research Articles

Most journal articles reporting results of original research include most or all of the following sections:

- a **Title page** including a self-explanatory title followed by a list of the author(s) and their affiliation(s) with contact information (usually e-mail address) for the *corresponding* author, that is, the one who will interact with the journal editor during the publication process,

- an **Abstract** which provides a condensed summary of the article including the nature of the problem addressed, the objective(s) of the research project, a brief description of the methodologies used, and a brief summary of the major research findings and their significance to the scientific community,

- an **Introduction** section which provides a literature review of the topic, identifies gaps in knowledge, and lists the objectives of the research reported,

- a **Methodology** section which provides a clear description of procedures used to accomplish each objective,

- a **Results** or **Results and Discussion** section which reports results obtained for each objective and discusses how each relates to the objectives of the project and to previous studies reported in the literature,

- an **Acknowledgments** statement (if applicable) to identify funding sources and/or to express appreciation to colleagues who provided substantial assistance during the project,

- a **Literature Cited** section which lists all references cited in the text in a specified order.

Different journals use different formats, and the best way to determine the appropriate format to use for a particular journal article is to read the "Instructions to Authors or Contributors" notice posted in a recent issue of the journal in question or on the journal's website. Instructions to authors indicate what sections are to be included in manuscripts and how they are to be labelled, and the format to use for literature citations. In addition, they specify procedures for submission of manuscripts and usually provide additional information on the journal's review and publication policies.

Preparation of Journal Articles

One of the best strategies for writing journal articles of original research is to begin with an outline containing the basic sections of the manuscript with headings and subheadings listed in logical order and containing specific information (Figure 7.1). Using a template such as this, you can compose the discussion of each topic in each section separately in any order desired (see Examples 7.1 and 7.2). When you tire of writing one topic or section, you can easily switch to another topic or section and resume writing. This is a very effective means of avoiding the irritating condition known as "writer's block" in which you know what you want to say but just can't come up with the right words. Eventually, the words will appear in your head, and switching to another topic or another section will keep you from losing a lot of valuable time while your brain is getting its act together.

When preparing each section of a journal research article, keep the following considerations in mind:

- the Abstract should be no longer than one paragraph and should provide a brief but comprehensive summary of the rationale, objectives, and results of the research project.

- the Introduction section should describe (1) the nature of the problem that was addressed, and why it is important to the scientific community and/or general public, (2) what is known about the problem, that is, a review of pertinent literature, (3) a brief discussion of what was not known about the problem (i.e., gaps in knowledge which existed) when the research was initiated, and (4) the objectives of the research (which should relate to one or more of the gaps in knowledge discussed in item 3). In most cases, the number of objectives listed should not exceed three or four.

- the Materials and Methods (or Methodology) section should clearly and concisely describe the procedures that were used to accomplish each of the objectives listed in the Introduction. Procedures should indicate when and where field and/or laboratory research was conducted, and how factors that may have influenced the results of field or laboratory experiments were addressed. For example, ". . . cohorts of plants were maintained in the laboratory under a constant temperature of 20°C, relative humidity of 70% and a 12:12: photoperiod under artificial lights . . ." All measurements should be based on the metric system, and the manufacturer and stated accuracy of each instrument used in measurements should be identified. For example,

Title
Author(s)
Affiliations
ABSTRACT
INTRODUCTION
- Nature of the Problem
- What is known — literature review
- Unknowns — gaps in knowledge
- Objectives

MATERIALS AND METHODS
- Objective 1
- Objective 2
- Objective 3

RESULTS (or RESULTS AND DISCUSSION)
- Objective 1
- Objective 2
- Objective 3

DISCUSSION
- Interpretation(s)
- Summary
- Additional research needs

ACKNOWLEDGMENTS
LITERATURE CITED

Figure 7.1 Format (template) for a typical journal article of original research.

". . . wavelengths of reflected energy were measured using a FieldSpec Pro® VNIR spectroradiometer (Analytical Spectral Devices, Boulder, CO) sensitive to wavelengths ranging from 350 nm (ultraviolet) to 1100 nm (near infrared) . . ." The Methodology section should be written with particular care because (1) it will be scrutinized by reviewers and editors to evaluate the validity of your research and (2) it will provide a "recipe" for other researchers wishing to replicate your research or to build on it in addressing related research problems.

- The Results section should summarize the results of studies or experiments for each objective separately without attempting to interpret such results (interpretations are normally discussed in the Discussion section). If combined into one section (i.e., Results and Discussion) both are usually discussed for each objective separately.

- The Summary or Conclusion statement is typically included as the last paragraph in the Discussion (or Results and Discussion section) and serves two useful purposes: (1) to consolidate the main ideas of the article into a clear "take-home message" for the reader and (2) to point

Example 7.1—Using the Template

MATERIALS AND METHODS

- Experiment 1.
- Experiment 2.
- **Experiment 3.** Data from experiment 3 will be analyzed using a Student's *t* test with pooled variances at the 5% probability level.

RESULTS (or RESULTS AND DISCUSSION)

- Experiment 1.
- Experiment 2.
- **Experiment 3.** The Student's *t*-test failed to detect differences among the means for experiment 3 ($t = 0.45$; $df = 14$; $P > 0.05$), which indicates that the compound does not appear to enhance growth in these plants.

Notice in this simple example that each subheading in the template represents a paragraph or more of text relating to that topic. You can simply work on any section or subheading you like, save your work, and switch to another topic if you begin to tire. By doing so, you will build a manuscript piece by piece and will probably not be burdened by writer's block.

Example 7.2—Using the Template (Continued)

MATERIALS AND METHODS

- **Experiment 1.** Plants were divided into 10 groups of 10 plants each and were maintained under constant lighting conditions (12:12 LD photoperiod) under laboratory conditions (describe conditions here). Five randomly selected groups were provided with the fertilizer compound (enter rate here) and the remainder served as controls.
- **Experiment 2.** It will work on this objective when I am tired of the other two.
- **Experiment 3.** Data from experiment 3 will be analyzed using a Student's *t*-test with pooled variances at the 5% probability level.

RESULTS (or RESULTS AND DISCUSSION)

- **Experiment 1.** It will enter results when data have been analyzed.
- **Experiment 2.** It will enter results when data have been analyzed.
- **Experiment 3.** The Student's *t*-test failed to detect differences among the means for experiment 3 ($t = 0.45$; $df = 14$; $P > 0.05$), which indicates that the compound does not appear to enhance growth in these plants.

Notice that the experimenter has written over the subheadings with preliminary results and notes of future plans. This process can continue until the manuscript has been completed. Note: if using this or any other procedure, be certain to save results routinely and back up your data.

out the need(s) for additional research on the problem addressed by the research. A summary statement also helps end the article in a smooth and graceful manner.

- The Acknowledgments statement expresses recognition for person(s) who contributed substantially to the success of the research project, but not to an extent that warrants co-authorship. Acknowledgment of support provided by funding sources is not only a matter of courtesy, but is generally required by the funding agencies themselves. In many cases, the funding agencies provide a "boilerplate" acknowledgment statement that should be included in both oral presentations and written manuscripts and other documents.

- The Literature Cited section should list all references cited in the text of the manuscript in the format specified by the journal (citation formats are discussed in Chapter 2). No reference should be listed in the Literature Cited section which is not also cited in the text of the manuscript—this must be verified by careful cross-checking.

Draft manuscripts should be reviewed carefully by the author(s) to ensure that there are no typographical errors and/or omissions, and that all authors are in agreement with the conclusions and recommendations included in the manuscript. Although not required by most academic institutions, many authors request *solicited reviews* of the manuscript from colleagues in their disciplines (this may be a requirement for scientists in the Federal and state governments, which also require administrative reviews of scholarly work). Moreover, it is advisable to screen all or portions of the manuscript with a plagiarism checker to ensure that all research discussed in the text has been properly cited (in this author's opinion, this should become a standard practice for all writers).

Other Research Reports

In addition to journal articles, results of original research may be reported in a variety of other written documents. Conference proceedings articles involving original research are similar in format to journal articles and may or may not be refereed. Laboratory reports similar in structure and format to journal articles are commonly included as requirements in university courses, but are usually based on studies or experiments of very limited scope and are rarely if ever published. Nevertheless, the value of classroom laboratory reports as training exercises for future scientists cannot be overemphasized. Other types of research reports that may or may not be published include periodic and final progress reports submitted by scientists to funding agencies and posted on websites such as the USDA CRIS system (see Chapter 11) and documents such as the Environmental Impact Statements published by the U.S. Department of Homeland Security and other government agencies as required by the National Environmental Policy Act. All of the information discussed in Part I of this textbook is applicable to journal articles as well as these other documents.

Theses and Dissertations

Theses are reports of original research conducted to satisfy university requirements for the Master of Science (M.S.) degree and *dissertations* serve a similar purpose for students seeking to satisfy requirements for the Doctor of Philosophy (Ph.D.) and equivalent degrees. The format for both documents varies considerably among universities, and both differ considerably from other types of primary literature. Theses and dissertations are reports of original research, but tend to be considerably longer (50–100+ pages) than standard journal articles and commonly contain sections (e.g., appendices) that are rarely if ever included in other types of primary literature. In many institutions, discussions of the research topic, background literature, methodology, and results of various studies are organized into separate chapters. In other cases, individual chapters are written in a format similar to that of standard journal articles. Theses and dissertations generally serve as the basis for one or more journal articles—they are not generally published in their original form, but most are copyrighted and posted in online databases such as Dissertations and Theses (http://search.proquest.com).

Summary

1. The *primary scientific literature* consists of articles from a variety of sources that report research findings that have not been published previously. Examples include theses and dissertations, unpublished laboratory reports, certain conference proceedings and journal research articles.

2. The format of journal research articles typically includes a title page, abstract, introduction, methodology section, results (or results and discussion) section(s), an acknowledgments statement and literature cited section.

3. The use of an outline formatted with headings and subheadings in a logical order can greatly facilitate writing journal articles and will go a long way in preventing writer's block.

4. Journal research articles and other types of primary literature should be written clearly and with as few words as possible, and all research conducted by scientists other than the author(s) must be cited in the text and in the literature cited section.

5. Articles published in refereed journals are generally considered to be more impactful than those published in nonrefereed journals.

Exercise 7.1

Select a few journals dealing with your field of expertise and check their ratings with a journal rating service (e.g., Journal Citation Reports (JCR): Science Edition http://thompsonreuters.com/en/products-services/schc). Compare ratings for several of the refereed vs nonrefereed journals.

Suggested Resources

Andersen, R. (1994). *Powerful writing skills.* New York: Fall River Press, 125 pp.

Hoffman, A. H. (2016). *Writing in the biological sciences: A comprehensive resource for scientific communication*, (2nd ed.). New York: Oxford University Press, 329 pp.

Hoffman, A. H. (2010). *Scientific writing and communication.* New York: Oxford University Press, 682 pp.

Knisely, K. (2013). *A student handbook for writing in biology,* (4th ed.). Sunderland, MA: Sinauer Associates, 318 pp.

Pechenik, J. A. (2013). *A short guide to writing about biology*. New York: Pearson, 276 pp.

Test Your Knowledge

Without referring to the material presented earlier in this chapter, mark an "x" in the box of the most appropriate answer to each question. Answers are provided at end of chapter.

1. Which of the following is an example of primary scientific literature?

❑ **a.** journal review article

❑ **b.** journal research article

❑ **c.** book chapter

❑ **d.** book

2. *Original* research is defined as research that ________________________.

❑ **a.** has not been published previously

❑ **b.** has been published in foreign countries but not in the United States

❑ **c.** is novel and highly innovative

❑ **d.** involves an extremely important topic

3. *True or False.* One of the best ways to avoid "writer's block" is to begin with the title page and write each subsequent section in sequence, thus avoiding the tendency to alternate from section to section.

❑ **a.** True

❑ **b.** False

4. *True or False.* The title of an article should be relatively short—no longer than about 16 words.

❑ **a.** True

❑ **b.** False

5. *True or False.* Abstracts of scientific papers should be brief—somewhere between 2 and 3 pages in length.

❑ **a.** True

❑ **b.** False

6. The statement of the problem addressed and objectives of the research are normally summarized in this section of a journal research article.

❑ **a.** Title page

❑ **b.** Introduction

❑ **c.** Methodology

❑ **d.** Results (or Results and Discussion)

7. *True or False*. In general, refereed journal articles are considered to be more impactive (important) that nonrefereed articles.

❑ **a.** True

❑ **b.** False

8. *True or False*. Theses and dissertations are most commonly reviews of previous research and are therefore considered to be secondary scientific literature.

❑ **a.** True

❑ **b.** False

9. *True or False*. In contrast to scientific articles published in most areas, those published in the United States and Great Britain use the English system of measurements.

❑ **a.** True

❑ **b.** False

10. *True or False*. Publications for research funded by Federal grants generally require an Acknowledgment statement identifying the funding source.

❑ **a.** True

❑ **b.** False

Answers: 1) b; 2) a; 3) b; 4) a; 5) b; 6) b; 7) a; 8) b; 9) b; 10) a.

Books, Book Chapters, and Review Articles

The Secondary Scientific Literature

Chapter Learning Objectives

After studying this chapter, you should understand:

- Why books, book chapters, and review articles are referred to as *secondary* scientific literature.
- The purpose of secondary scientific literature, and the important role it plays in the dissemination of scientific knowledge.
- The format, content, and procedures involved in the publication of secondary scientific literature.

The *secondary scientific literature* consists of articles from a variety of sources that report research findings that have been published previously. The major categories of secondary scientific literature include the following:

- **Journal Review Articles**—these are summaries and critiques of contemporary literature in specific topic areas. Most major journals publish both research articles and reviews in the same issues, although many journals are dedicated entirely to reviews (Table 8.1).

Table 8.1. Selected refereed journals dedicated to review articles.

Biology	
Annual Research and Review In Biology	www.sciencedomain.org/journal
The Quarterly Review of Biology	www.uchicago.edy/toc
Chemistry	
Review Journal of Chemistry	www.springer.com/chemistry/journal
Chemical Reviews (ACS)	www.pubs.acs.org/journal
Physics	
Reviews in Physics	www.journals.elsevier.com/reviews
Reviews of Modern Physics	www.journals.aps.org/rmp

In contrast to the format used in reports of original research (Chapter 7), review articles typically consist of the following sections (Figure 8.1):

- **Title and author affiliations**—similar to those for research articles,
- **Abstract**—similar to those for research articles except that discussion pertains to topics other than experimental results,
- **Topic Headings and Subheadings**—these summarize past and present research pertaining to each major and minor topic, and are placed in a logical sequence that will lead the reader to a conclusion or take-home message,
- **Summary**—optional
- **Acknowledgment**—included if applicable
- **Literature Cited**—includes all citations in text in the style preferred by the journal.

Notice that the headings and subheadings in the review article refer to topics rather than procedures and results. All of the writing principles discussed in Parts I and II are applicable to journal review articles, and a template similar to the one used for research articles (Chapter 7) can be used to develop review articles section by section as discussed previously.

- **Books and Book Chapters**—books are produced for a variety of purposes and the chapters they contain are written in a variety of formats. Two basic forms of books include those in which

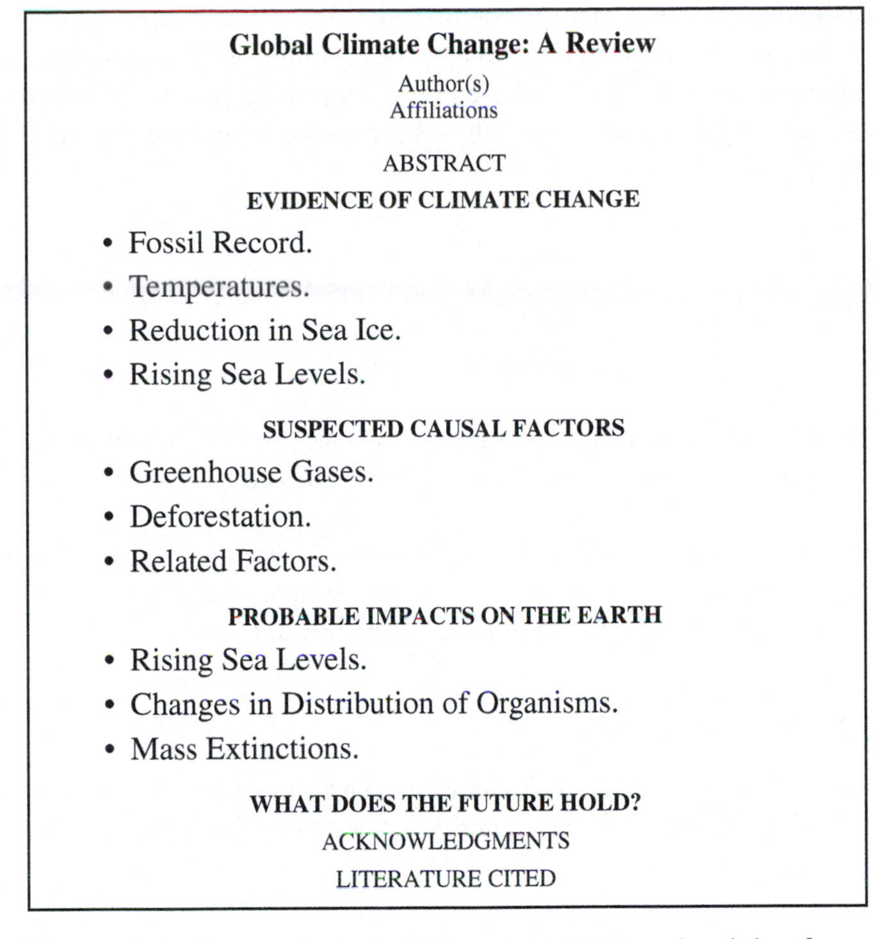

Global Climate Change: A Review

Author(s)
Affiliations

ABSTRACT

EVIDENCE OF CLIMATE CHANGE

- Fossil Record.
- Temperatures.
- Reduction in Sea Ice.
- Rising Sea Levels.

SUSPECTED CAUSAL FACTORS

- Greenhouse Gases.
- Deforestation.
- Related Factors.

PROBABLE IMPACTS ON THE EARTH

- Rising Sea Levels.
- Changes in Distribution of Organisms.
- Mass Extinctions.

WHAT DOES THE FUTURE HOLD?

ACKNOWLEDGMENTS

LITERATURE CITED

Figure 8.1. Format (template) for a typical journal article of original research.

- one or more authors participate in writing a series of chapters that are compiled into a single publication,
- one or more authors participate in writing each of a series of chapters that are edited by the same or different individuals into a single publication (edited book) that usually addresses a specific topic or theme.

Most college-level textbooks and the book you are reading now are examples of the first type. When citing this type of book, the entire book is considered a single publication and is cited as such, for example:

Smith, RI, Jones, BJ, Gutierrez, JM. 1965. Hypothetical Ecology. J&W Publishers, New York. 580 pp.

An example of the second type (edited book) is a classic text edited by C. B. Huffaker and P. S. Messenger (1976) which contained six sections and a total of 28 papers dealing with the biological control of insect pests. In an edited book, each individual chapter is considered a separate publication and is cited as such, for example:

Wilson, F., and C. B. Huffaker. 1976. The philosophy, scope and importance of biological control. Pp. 3–14. In Theory and Practice of Biological Control (C. B. Huffaker and P. S. Messenger, eds.). Academic Press, New York. 788 pp.

- **Annual Reviews and Monographs—**Annual Reviews (www.annualreviews.org) and similar publications provide authoritative reviews of literature for the past several years on given topics. Monographs are lengthy treatises on a topic area and are commonly published as monographic series, an example of which is Advances in Experimental Medicine and Biology (www.worldcat.org).

Summary

1. The *secondary scientific literature* consists of articles from a variety of sources that report research findings that have been published previously.

2. This category of the scientific literature serves to provide information regarding recent advances and publications to scientists and others involved in research.

3. Formats for different categories of secondary literature are diverse, but determining which to use in a particular journal submission is relatively simple—check the Instructions to Authors on the journal's web site or in a recent issue of the journal.

References

Wilson, F., & Huffaker, C.B. (1976). The philosophy, scope and importance of biological control. In C. B. Huffaker, & P. S. Messenger (Eds.), *Theory and Practice of Biological Control* (pp. 3–14). New York: Academic Press. 788 pp.

Test Your Knowledge

Without referring to the material presented earlier in this chapter, mark an "x" in the box of the most appropriate answer to each question. Answers are provided at end of chapter.

1. All of the following are examples of *secondary scientific literature* EXCEPT.
   - ❑ **a.** book chapters
   - ❑ **b.** journal research articles
   - ❑ **c.** journal review articles
   - ❑ **d.** college-level textbooks

2. *True or False*. Aside from format, the writing principles for primary and secondary literature are essentially the same.
   - ❑ **a.** True
   - ❑ **b.** False

3. *True or False*. In college-level textbooks written by multiple authors, each chapter is considered a separate publication for the author(s) involved in that section of the text.
   - ❑ **a.** True
   - ❑ **b.** False

4. *True or False*. In professional textbooks written by multiple authors and edited by one or more individuals, each chapter is considered a separate publication for the author(s) involved in that section of the text.
   - ❑ **a.** True
   - ❑ **b.** False

5. *True or False*. In contrast to primary scientific articles, secondary literature is rarely if ever subjected to anonymous peer review.
   - ❑ **a.** True
   - ❑ **b.** False

Answers: 1) b; 2) b; 3) b; 4) a; 5) b.

The Publication Process for Scientific Literature

From Draft to Published Work

Chapter Learning Objectives

After studying this chapter, you should understand:

- The normal sequence of events that occurs during the publication process for journal research articles and other scientific documents.
- The purpose and importance of *anonymous peer reviews,* and how they are conducted.
- How an author should properly respond to reviews of manuscripts provided by anonymous reviewers and editors of scientific journals.
- The purpose of a *galley proof* why it is absolutely critical for the author(s) to *proofread* it carefully prior to its publication.

Once a study or experiment has been completed and the data have been analyzed and interpreted (or the literature for a review article has been assembled), the author(s) normally develop a *draft* manuscript which may be sent to colleagues with requests for their comments and suggestions (these are referred to as *solicited* reviews are optional in most cases). Once the manuscript has been completed and reviewed to the author(s)' satisfaction, it is normally submitted to an editor of the journal or other source where it is to be published. In this chapter, we review the sequence of events that occurs from the time a manuscript is received by the editor of a *refereed* journal (one in which submitted manuscripts are subjected to anonymous peer reviews—see following section) until its final printing in a conventional journal or posting in an online journal (due to high publication costs, online journals are becoming increasingly commonplace).

Journal Research Articles

In most journals, information regarding manuscript formatting and submission procedures are provided in an *Instructions to Authors* statement posted on the journal's website or printed in the journal itself. Submission procedures vary from providing either (1) a specified number of paper copies or (2) an electronic copy of the manuscript (usually pdf) submitted online. When an editor receives the manuscript, he usually notifies the corresponding author and provides information regarding the journal's review policies and instructions for queries regarding the manuscript.

Anonymous Peer Reviews. If the journal is *refereed,* the editor will then forward printed or electronic copies of the manuscript and one or more review forms (Figure 9.1) to a specified number of anonymous reviewers who are experts in that particular field and whose identities are not revealed to the author(s) of the manuscript. For many years, this process has provided the principal means of maintaining quality in scientific publications.

Reviewers are typically asked to rate the manuscript based on any or all of the following factors, including

- **significance and relevance of the research**—the research should address a problem of significance, should contribute new knowledge, and the topic itself should be consistent with the mission of the journal,

- **the background literature**—the background literature included in the manuscript should include most or all of the major research relevant to the topic, and should support the need for research on the objectives reported,

- **soundness of the experimental design and methodology**—the experimental design should be sound, the sampling procedures should be unbiased and sample sizes should be adequate, and statistical tests should be appropriate and be interpreted correctly,

- **journalistic quality**—the manuscript should be well-organized, clearly written, and be grammatically correct and understandable,

- **other factors related to quality—**the manuscript should contain no evidence of plagiarism or any other form of academic misconduct.

Journal of Hypothetical Science
Manuscript Review Form

Instructions to Reviewers: Please review the manuscript indicated below and submit your recommendations to the journal editor (jhs@hsu.edu) within a period of 14 days from the date of receipt. Indicate the reviewer number you were assigned—Please do not enter your name anywhere on the review form. If you are unable to review the manuscript at this time, please contact editor so that another reviewer may be selected.

Manuscript # ________________ Reviewer________________

Recommendations

Indicate choice by entering "x" in designated space:

1) The objectives are stated clearly. __Yes ___No
2) Topics are discussed clearly and in logical sequence. __Yes ___No
3) Factual information discussed is correct. ___Yes ___No
4) Statistical design is sound. ___Yes ___No
5) Interpretation is based on adequate data. ___Yes ___No

Based on my review, manuscript should be:

___Accepted in present form
___Accepted with revisions
___Rejected with option to resubmit at later time
___Rejected with no option to resubmit

Specific Comments (if additional space is needed, please use additional page).
Page # Line # Comment

Confidential Comments to Editor:

Figure 9.1 Typical manuscript review form sent to anonymous reviewers by editors of refereed journals.

When reviewing manuscripts, an anonymous reviewer should remember that he or she has the dual responsibilities of (1) helping fellow scientists improve the quality of their publications, while (2) protecting the integrity of a journal or publisher. When reviewing manuscripts that have deficiencies but appear to be salvageable, reviewers should

- provide constructive suggestions to the author(s) for improving the organization and flow of information in the manuscripts,
- point out any major studies that should have been addressed in the manuscript but were not,

- suggest alternative statistical tests if the ones used in the manuscript were inappropriate,
- avoid rejecting manuscripts on trivial grounds, and never be rude or condescending in anonymous reviews.

On the other hand, the reviewer has an obligation to the journal editor to reject outright any manuscript that

- fails to discuss a significant number of important papers dealing with the topic in question,
- was so poorly designed that the data are invalid,
- shows evidence of data fabrication, data falsification, and/or plagiarism.

After reviewing the manuscript, the reviewers then make recommendations ranging from (1) "reject" to (2) "accept with revisions" to (3) "accept in present form" (Figure 9.1). The criteria for rejection or acceptance are usually based on whether or not the research is scientifically sound and the manuscript is well-written and grammatically correct.

Once the editor receives recommendations of the anonymous reviewers, the original manuscript and review forms (without reviewer's names) are returned to the corresponding author with instructions of how to proceed. On most review forms, there is also a provision for "Confidential Comments to Editor" which is removed from the copy sent to the authors—this allows reviewers to express concerns relating to issues that the editors should be aware of (e.g., suspected plagiarism or other misconduct).

Responding to Reviews and Editorial Comments. One important point that all writers should understand is that comments by reviewers and editors are generally intended to be constructive and helpful, and are rarely considered the "final verdict" on how a scientific manuscript should be written. If an author feels that one or more comments by reviewers are unwarranted or irrelevant, he or she should contact the editor and respond to the review(s) in a logical and polite manner. In cases in which the writer presents a strong and valid defense of his methodology and interpretations, most editors will agree and approve the manuscript (or parts thereof) as originally written.

However, writers should also understand that seemingly adverse reviewer's comments are an indication that one or more colleagues in his or her discipline had problems understanding the content of the article and/or disagreed with the interpretation(s) for some reason. In both cases, writers should make every attempt to address legitimate concerns by colleagues and make necessary revisions before manuscripts or other documents are published. In no case should writers "fly off the handle" or brood for days after receiving what they consider to be adverse comments by reviewers—the peer review process has been used for centuries to maintain quality of scientific publications, and it has worked well for the most part. If a revision is required, it will usually improve the quality of the published manuscript.

Preparation of a Revised Manuscript. If a manuscript requires one or more revisions, the author has the responsibility of addressing each of the required revisions. In many cases, the revision involves correcting errors in spelling, syntax, sentence structure, and other minor changes. In other cases, the reviewer (and editor) may require major revisions in the form of rewriting a major part of the manuscript to improve clarity. A description of each change made should be summarized in a letter addressed to the editor, and the revised manuscript should be returned to the editor.

Example 9.1 Responding to Reviewer's Comments

Following two examples illustrate proper and improper ways to respond to a reviewer's comment that a particular statistical test used in the author's manuscript was invalid and that another test should be used instead.

Reviewer Comment:

"Data of the type summarized in Table 3 are normally analyzed using one of the nonparametric tests based on chi-square. These data should be reanalyzed before the paper is published in this journal."

Improper Response:

Reviewer 2 suggested that we use a nonparametric test rather than the ANOVA we reported. This is nonsense! If Reviewer 2 had bothered to take a closer look at our data, he or she would have seen that the use of ANOVA was clearly justified for this experiment. We request that you ignore this comment by Reviewer 2, and also suggest that you assign manuscripts to more competent reviewers in the future.

This comment is highly confrontational and is likely to offend the editor (who may be a good friend or relative of the reviewer—or may have been the one who reviewed the paper in the first place) and result in a rejection of the manuscript, even though the test used in the experiment was valid.

Proper Response:

Regarding the comment by Reviewer 2 that we should have used a nonparametric test instead an ANOVA in Experiment 2, please note that our sample size was large ($n = 60$ observations per treatment) and our data satisfied all of the assumptions and conditions for use of parametric tests such as ANOVA. Although we could have used the test suggested by Reviewer 2, the use of ANOVA is clearly justified and we prefer to publish the manuscript using this test.

The author effectively refutes the comments by Reviewer 2 **tactfully and politely** and requests that the editor approve of the author's use of the ANOVA in this experiment. The editor will most likely approve the author's request.

Preparation and Review of a Galley Proof. Once an editor has accepted the revision, the manuscript is then submitted to a publisher (of printed journals) or to a managing editor who prepares an electronic file of the revised manuscript which will be posted in an online journal. In both cases, a "galley" proof of the revised manuscript will be returned to the corresponding author for approval before it is printed or posted online. At this point, it is critical that the author(s) of the manuscript proofread the galley to ensure that it has been prepared correctly and is ready for publication. If any changes in the galley are necessary, the author can indicate these by using *proofreader's marks* examples are available on numerous web sites, e.g., (www.galenahigh.com) or by summarizing these in an e-mail message (preferably both). Once the editor and/or publisher receive the author(s) approval, the manuscript is published or posted online and thus becomes a permanent document that is available to the scientific community and the world.

Publication of Books and Related Documents

The publication process for books and related documents is variable depending on the publisher. Publications by government agencies are almost always subjected to internal reviews by agency administrators, and may be sent to outside reviewers. Once published by the Government Printing Office (GPO) they may be accessed through a number of online databases such as *GPO Access, AGRICOLA,* and others. Commercial publishers of scientific books most commonly send copies of books to selected reviewers whose comments may be posted in review sections of journals and other professional publications (Murphy 2016). Titles of recent books are readily available in online databases such as *Books in Print* and others.

Summary

1. In most journals, information regarding manuscript formatting and submission procedures are provided in an Instructions to Authors statement posted on the journal's website or in a recent issue of the journal itself.

2. Manuscripts submitted to refereed journals are subjected to anonymous peer reviews which have long been used to maintain quality in scientific publications.

3. Authors should regard comments by anonymous reviewers and editors in a constructive manner, although comments that the authors regard as incorrect or unwarranted can be challenged if done properly.

4. Galley proofs sent to authors by the publisher should be proofread carefully and thoroughly, and any errors should be reported to the publisher prior to publication. Once published, manuscripts become part of the scientific literature and are available to the scientific community and the world.

References

Murphy, S. (2016, March 1).The Review Process: How a Book Gets Reviewed. http://www.writing-world.com/promotion/murphy1.shtml

Test Your Knowledge

Without referring to the material presented earlier in this chapter, mark an "x" in the box of the most appropriate answer to each question. Answers are provided at end of chapter.

1. What is the primary purpose of anonymous peer reviews of manuscripts submitted for publication in scientific journals?

❑ **a.** to maintain quality of scientific publications

❑ **b.** to satisfy regulations mandated by Federal and state funding sources

❑ **c.** to assign "impact factors" to an author's publications

❑ **d.** to prevent copyright violations by other scientists

2. When conducting anonymous peer reviews, reviewers typically scrutinize all of the following factors EXCEPT

❑ **a.** integrity of the experimental design

❑ **b.** statistical methodology used to analyze data

❑ **c.** significance and relevance of the research problem addressed

❑ **d.** the country or region of the world in which the research was conducted

3. All of the following are considered legitimate reasons for rejecting a manuscript EXCEPT

❑ **a.** awkward writing style

❑ **b.** failure to include major articles dealing with the subject of the research

❑ **c.** major flaws in experimental design

❑ **d.** insufficient sample size(s) in surveys and experiments

4. *True or False*. Once an author receives comments from anonymous reviewers and the editor's comments, the author(s) have little choice but to revise the manuscript accordingly or withdraw the paper.

❑ **a.** True

❑ **b.** False

5. Once the author(s) have returned a revised manuscript to the editor and it has been accepted for publication, what is the next step in the publication process?

❑ **a.** preparation of a galley proof

❑ **b.** publication of the revision accepted by the editor

❑ **c.** revised manuscript is resubmitted to anonymous reviewers

❑ **d.** none of the above

Answers: 1) a; 2) d; 3) a; 4) b; 5) a.

Important Documents

Letters of Application, *Curriculum Vitae*, and Personal Statements

Documents for Job Applications

Chapter Learning Objectives

After studying the material in this chapter, you should understand:

- How admission and job vacancy announcements are developed and posted by academic and research institutions.
- How to prepare and submit acceptable Letters of Application for job announcements.
- How to develop effective *Curriculum Vitae* (CVs) documenting your education, training and professional accomplishments.
- How to develop effective Personal Statements summarizing your values, career goals and contributions you would like to make to your chosen profession.
- Important considerations in the preparation and submission of admission and job application documents that may make the difference between success and failure.

Sometime before their anticipated graduation dates, undergraduate students across the country will begin the process of applying for admission to graduate programs in various academic institutions while others will begin applying for jobs in the public and private sectors. They will be in the good company of recent graduates with advanced degrees seeking their first jobs and employed scientists and educators seeking new positions for career advancement purposes. How successful these efforts prove to be will depend to a great extent on the quality and contents of several documents that are normally included in admission and job application packages, including those submitted electronically.

A typical vacancy announcement for a professional position in a hypothetical academic institution is shown in Figure 10.1. Notice that the announcement includes (1) the name and physical address of the entity posting the announcement (Department of Biology, Hypothetical University, (2) the title of the position (Biological Technician) and its job vacancy number (F13/12-014), and (3) the minimum qualifications for employment (in this case, 12 hours of courses in the biological sciences or related disciplines) and other factors that are preferable but not mandatory for hiring (previous laboratory experience). The announcement also (1) summarizes duties of the position, (2) provides the name(s) and contact information for the intended recipient(s) of the application, and (3) indicates the deadline for receipt of the application materials by the hiring entity. In order to comply with federal and state Equal Employment Opportunity (EEO) guidelines, the ad will be posted in one or more public media (e.g., Academic Careers Online at http://www.academiccareers.com/Employers.html) for a specified period of time before the interview and hiring process begins.

Letters of Application

A letter of application (or cover letter) is typically the first document included in a conventional (hardcopy) application package, or may be included as an attachment for applications submitted online. The letter of application should provide reviewers of the application with the following critical

Position Announcement
Biological Technician

The Department of Biology in Hypothetical University of South Texas seeks to fill a Biological Technician position beginning as soon as position is filled (Job Vacancy F13/12-014). A minimum of 12 hours of college level Biology courses is required, and previous laboratory experience is preferred. The selected candidate will be expected to assist faculty in the development of laboratory experiments for use in general biology courses, and to maintain a safe working environment for undergraduate and graduate students. Information on the HU Biology Department can be found at www.hu.edu/dept/biology .

Applications should include a letter of application, CV, Personal Statement of the applicant's background and qualifications for this particular position and three (3) letters of reference. Applications and letters of reference should be addressed to: Biology Search Committee, Hypothetical University, College of Science, 1201 West Hypothetical Drive, Somewhere, TX 79999. Due date for applications is Friday, February 20, 20XX. Incomplete or late applications will not be considered. Review of applications will begin on Wednesday, February 25, 20XX.

UH is an Affirmative Action/Equal Opportunity employer. Women, minorities, and qualified individuals are encouraged to apply.

Figure 10.1 Position announcement posted by hypothetical academic institution.

information: (1) the applicant's contact information, including home address, phone number(s) and e-mail address(es), (2) the specific program or position that the applicant is applying for including its job vacancy or announcement number, and (3) one or more statements verifying that the applicant meets or exceeds the minimum requirements for the position in question. The letter should also convey a message that the applicant understands the scope and requirements of the position and is enthusiastic about the prospect of being granted an interview. In the final portion of the letter, the applicant should thank the reviewers for considering his or her application, and indicate that their curriculum vitae and personal statements are attached. Letters of application should be brief (1 ½–2 pages maximum) and should be signed by the applicant in blue or black ink.

A sample letter of application responding to this announcement is shown in Figure 10.2. Notice that the letter is brief and to the point. In the first two sentences, the applicant clearly indicates the

1

John Q. Public
510 E. Commerce, Somewhere, TX 98999
Jqp@aol.com

February 1, 20XX

2

Department of Biology, Hypothetical State University
1010 Imaginary Drive
Anywhere, TX 77410

To the Biology Search Committee:

3

I read with interest the position announcement for a Biological Technician (Job Vacancy F13/12-014) and am interested in applying for the position. I am currently attending UTPA as a Biology major and Chemistry minor, and have completed 28 hours in Biology and 12 hours in Chemistry. During my undergraduate studies in the UTPA Biology Department, I have served as a Laboratory Teaching Assistant (4 semesters) and have gained considerable experience in the use of analytical tools used in molecular biology and biotechnology. My anticipated date of graduation is May, 2016 after which I am planning to continue my education in an M.S. program at UTPA or another institution.

4

I am very interested in the Biological Technician position that was recently advertised as it will help me launch a career in the biological sciences. I am currently enrolled in several classes but will be available at your convenience if you call me for an interview. Thank you very much for considering my application, and my CV and Personal Statement are attached.

5

Sincerely,

John Q. Public

Attachments (2)

Figure 10.2 Letter of application for hypothetical job announcement. Applicant's name is bolded and centered at top of page along with contact information (1). Letter is dated and addressed to search committee as per instructions in vacancy announcement (2). Applicant indicates the specific position he is applying for (Biological Technician) and its position vacancy number (F13/12-014), documents that he meets educational requirements (28 hours of Biology and 12 hours of Chemistry), and has preferred qualifications (experience as lab teaching assistance for four semesters) (3). Applicant indicates interest in position and enthusiasm over the prospect of obtaining an interview (4). Applicant thanks search committee for considering his application and indicates that his CV and Personal Statement are attached (5).

title and vacancy number for the position he is applying for and notes that his educational level (28 hours in biology and 12 hours in Chemistry) exceeds the minimum requirements listed in the position announcement. At the end of the first paragraph, he lists experience and skills that are clearly related to the duties of the position (employment as a lab teaching assistant for four semesters) and in the last paragraph, he provides convincing evidence that he is genuinely interested in the position because of its potential to launch his scientific career. He concludes by thanking the search committee for considering his application and indicates that his CV and Personal Statement are enclosed.

Curriculum Vitae

A *Curriculum Vitae* or CV (sometimes referred to as a resumé) is a document that summarizes a person's educational background, work experience, scholarly contributions, awards and honors, and other accomplishments made during his or her educational and professional career. CVs are required in virtually all applications for admission to academic programs and for employment in professional positions. They are also one of the primary means by which search committees determine whether or not a given applicant meets the minimum requirements for the advertised position, and how he or she competes with others in the applicant pool for the position. Although there are currently no mandated formats for CVs, certain designs tend to be more effective than others for a variety of reasons.

Example 10.3

A CV for a recent graduate of a hypothetical academic institution is shown in Figure 10.3. The CV provides the person's full name and contact information (home address, phone number(s) and e-mail address) centered at the top of the first page directly below the heading "Curriculum Vitae." The person's name is bolded and printed in a slightly larger font that the contact information in order to make it conspicuous and draw the reader's attention to it. Along the left margin, a series of headings (in bold font) summarize the person's educational history, work experience, scholarly contributions, certifications, awards and honors, and any other factors that may be of importance in obtaining employment or advancing in a current job.

Because of its importance in determining whether or not a given applicant meets the minimum educational requirements for an advertised position or warrants promotion in a current position, the heading "Education" should always be located near the beginning of a CV (Figure 10.3). One of the most common conventions is to list degrees conferred by academic institutions in order of rank, that is, doctoral degrees (which may be abbreviated as Ph.D., M.D. or Sc.D.) are listed first, followed by Master's degrees (M.S. or M.A.), Baccalaureate degrees (B.S. or B.A.) and Associate degrees. Degrees of equivalent rank are most commonly listed in descending chronological order, that is, the most recently conferred degree is listed first and the earliest conferred is listed last. In all cases, the degree listed should be followed by the name of the institution, the field or discipline of study, and date on which the degree was conferred. If the degree is near completion but has not yet been conferred, the anticipated date of graduation should be indicated (see Figure 10.3). This is an important consideration for students applying for admission to programs or jobs before graduation—in many if not most cases, applicants who do not meet the minimum educational requirements on technical grounds may be considered eligible for hiring if it appears imminent that the required degree will be conferred on or before the official starting date of the position.

Entries under the remaining headings are usually listed in similar manner. Scholarly contributions are most commonly listed under the separate headings of "Presentations" and "Publications" using an accepted citation format (Figure 10.3). Although citation formats vary among journals, a single

Curriculum Vitae

John Q. Public

510 E. Commerce, Somewhere, TX 79999

johnqpublic@aal.com

Education

B.S., Hypothetical University, major Biology, minor Chemistry, May 20XX.

Work Experience

Biological Technician, Department of Biology, Hypothetical University, 20XX.

Presentations

Public, John Q. "Biology of Species X," Annual Meeting of the Society of Science, Anywhere, TX, 4-8 May 20XX.

Publications

Public, John Q. 20XX. Biology of Species X in Texas. J. Soc. Science 41:55-60.

Membership in Professional Societies

Society of Science

Honors and Awards

Dean's List, Hypothetical University, Fall, 20XX.

Special Skills

Computer programming

Extracurricular Activities

Men's softball team, Hypothetical University, 20XX.

Personal References

Available on Request

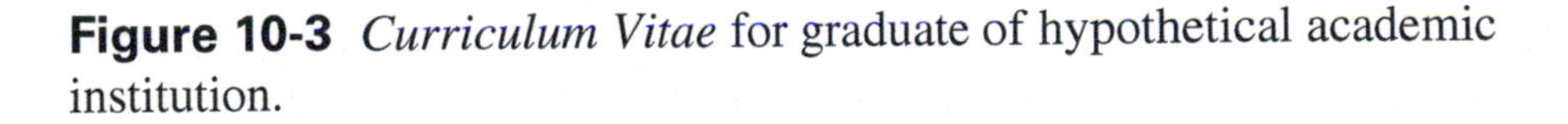

Figure 10-3 *Curriculum Vitae* for graduate of hypothetical academic institution.

acceptable format should be used for all presentations and publications included in the CV. Citations for articles published in journals should always include a list of the author(s), the year of publication, the title of the article, followed by the journal name, volume number and page numbers of the article. Citations for oral presentations are basically similar except that the title is usually enclosed by quotation marks, and the title of the meeting or conference, the sponsoring society or organization, and date(s) of the meeting are listed. Although a variety of citation formats exist and are acceptable (see Chapters 2, 4, and 7), the particular format used should be consistent throughout the CV.

Undergraduate and graduate students (and even some new faculty members in academic and research institutions) would be well advised to develop a "master CV" documenting your previous training and professional accomplishments, and to update it routinely as your career progresses. Trying to keep track of your professional accomplishments by filing printed documents and written notes in manila folders is inefficient and fraught with hazards—paper documents have an uncanny way of becoming misplaced or disappearing forever. As your career progresses, your master CV will appear to grow exponentially and (somewhat surprisingly) will never be used in job application packages because of its size. However, because it contains most or all of your significant professional accomplishments, the master CV provides an ideal mechanism by which to generate "custom CVs" (sometimes referred to as resumés) for specific job announcements or other needs. If the master CV is well-organized and comprehensive, custom CVs can be generated in a fraction of the time that would have been required otherwise.

Personal Statement

In contrast to the CV, which is usually written in a bullet-style format, the Personal Statement is written in narrative form and provides the reviewer(s) of the document with additional information about the applicant that is not normally included in his or her CV. A Personal Statement should provide a brief summary of the applicant's history, values and interests, career goals and contributions that he or she would like to make to their chosen profession. In applications for faculty positions involving a combination of teaching and research, the single Personal Statement may be replaced by separate documents for Teaching Interests and Research Interests. In all cases, the Personal Statement (or its variations) should be prepared with great care as it provides reviewers with considerable insight into the applicant's attitudes and values on certain issues that may affect job performance, and concurrently provides a free writing sample by which to judge the applicant's written communication skills.

Example 10.2

A Personal Statement included in an application package of a hypothetical student applying for admission to a medical school program is shown in Figure 10.4. Notice that the student begins by explaining that he became interested in the natural sciences at a very early age, and that interest was intensified during his high school years when one of his instructors helped him treat a bird with a broken wing. The satisfaction of successfully treating a bird eventually evolved into an interest in treating humans and was instrumental in his decision to study medicine (this statement very clearly explains his desire to become a medical doctor). He goes on to mention that his undergraduate studies in Pre-Med Biology were difficult due to financial difficulties and several family tragedies, but he was able to complete his studies because of his determination to become a doctor and his strong religious faith. Religion is a sensitive subject (remember that many religions are antagonistic toward other religions, and some people in administrative positions are agnostics or atheists), but this statement was written in a manner that clearly indicated strong religious values of the writer but also indicated no tendencies to become a crusader if hired (which might be considered a "red flag" by some potential employers). Finally, the writer indicates that his intent is to study orthopedics (surgical correction of skeletal deformities or injuries, and study of bone and muscle diseases) and to open a practice in an area that is short of specialists in this discipline. Thus, this brief personal statement explains why the person developed an interest in medicine, what he intends to specialize in if he becomes a medical doctor, and the contributions he intends to make to the field of orthopedics if he is successful in completing medical school and opening a practice.

Job announcements for faculty positions at academic institutions may specify statements of "Research Interests" and "Teaching Interests" in lieu of the Personal Statement. In the statement on Research Interests, the applicant should discuss the scope and importance of his current and past research, and the type of research program that he or she would like to develop if hired by the institution. The proposed research program should be consistent with duties listed in the job announcement and the discussion of research interests should indicate how such a program will benefit the hiring institution and its students, including their participation and training in research projects. In the statement on Teaching Interests, the applicant should discuss his or her previous

John Q. Public

Personal Statement

I became interested in the biological sciences at an early age and developed a fascination for birds. As I was growing up, I learned identify most of the birds that occurred in my home state and had a variety of avian pets, including two parrots, ten parakeets and several pigeons and doves. I tried to buy a 150-pound flightless emu but my parents refused.

A particularly significant event in my life occurred while I was in high school. My biology teacher was aware of my interest in birds and contacted me one day and asked if I would help treat a dove he had found with a broken wing. We patched the bird up, put splints on its broken wing, and held it in the lab for about three weeks. The bird had recovered by then and I had an enormous sense of accomplishment when I watched it fly off into the trees.

This event was pivotal and truly changed my career plans. If I can heal birds, why not do the same for humans? I decided to pursue a career in medicine and enrolled in the pre-Med program at Hypothetical University. My undergraduate studies were stressful because of severe financial difficulties compounded by the death of my parents in a car accident, but I succeeded because of my determination to become a doctor and my strong religious faith.

If I succeed in medical school, I intend to specialize in orthopedics and would like to open a practice in my hometown to fill a major void in the availability of health care, as the closest orthopedic surgeon is located nearly 200 miles away.

Figure 10.4 Personal Statement by graduate of hypothetical institution.

experience in teaching activities, including teaching assistantships, post-doctoral assignments in teaching, and previous faculty positions. In addition, the applicant should indicate which of the current courses offered by the hiring institution he or she is qualified to teach, and new courses that could be developed and taught if the applicant is hired. The applicant's philosophy on teaching methodology, grading procedures, and interactions with students inside and outside of the classroom should be discussed in detail.

Important Considerations for Job Application Documents

The documents included in application packages are fundamentally important in obtaining new employment and advancing in a current position. Important considerations in the preparation and submission of these documents include the following:

- Letters of Application and CVs both contain the applicant's name and contact information. The applicant's given name should be spelled out in full, and nicknames should be avoided. The e-mail address should be professional and conventional—avoid exotic or provocative addresses such as bandito777@aol.com or hawghunter@aol.com.

- In Letters of Application, always indicate the specific job title and job announcement number you are applying for. Search committees may be handing multiple announcements and are usually on very tight schedules. Because of potential legal issues, they cannot and will not try to guess what position you are applying for—if there is any doubt about what position you are applying for, your application will almost certainly be rejected.

- Letters of Application and Personal Statements should be relatively brief (1½–2 pages maximum) and to the point. Anything longer will probably not be read.

- Statements in Letters of Application and Personal Statements should reflect positively on the applicant—avoid calling attention to any real or perceived weaknesses of yourself. For example, the statement ". . . because of my difficulties in reading, I usually have to read scientific articles multiple times before I can understand them . . ." will suggest learning disabilities to most reviewers, while the statement ". . . I generally read scientific articles several times in order to ensure that I understand the author's message completely . . . " reflects positively on the applicant without stretching the truth.

- Headings and entries included in CVs should be consistent with the applicant's experience at the time the CV is written. If an undergraduate student has not progressed to the point where he or she has given oral presentations at conferences or published in refereed journals, those headings should be omitted from the CV. Include only headings that are appropriate and avoid any headings in which accomplishments would be "None" or "Not Applicable."

- Deadlines published in the position announcement generally refer to the date on which the application is received by the hiring institution (not to when it is submitted). For legal reasons, most institutions will not consider applications received after the published deadline unless it is officially extended for all applicants by the institution itself.

- If a position announcement indicates that letters of reference by colleagues should be sent to the search committee, make certain that the colleague knows the deadline and complies. If any part of an application is missing (including letters of reference) at the time of review, the entire application will usually be rejected.

- First impressions are extremely important in job applications, and neatness is imperative. All documents with multiple pages should be stapled together neatly, and separate documents included in the application page should be attached using paper clips. Reprint any pages of any documents with smudges or stains. This may sound nit-picky, but sloppiness in the preparation of application documents may be interpreted as potential for sloppiness on the job.

Exercise 10.1

Develop a master CV that summarizes your past and present accomplishments. Include headings that are appropriate for your stage of career development and enter accomplishments that are appropriate for each; for example, under "Honors and Awards," list each semester that you have been included on the Dean's List or any memberships in Honor Societies, etc. Under "Education," list the degree you are working on and the anticipated date of graduation. Begin to update your master CV as your accomplishments begin to accumulate.

Exercise 10.2

If you already have a large CV or can get an electronic copy from someone who does, practice developing a custom CV by cutting and pasting headings and entries to match this requirement: Custom CV for this exercise should include education level, work experience during past two years, scientific contributions or membership in professional societies during the past year. Maximum length—two pages. Time yourself on how long it takes to generate this custom CV.

References

Lewis, A. (1993). *Best resumes for scientists and engineers.* New York: John Wiley & Sons, 224 pp.

Test Your Knowledge

Without referring to the material presented earlier in this chapter, mark an "x" in the box of the most appropriate answer to each question. Answers are provided at end of chapter.

1. *True or False*. If you are applying for two or more positions that have been advertised by an agency or institution, it is considered appropriate to combine them into one pooled application in order to save time and reduce paperwork.

   ❑ **a.** True

   ❑ **b.** False

2. *True or False*. Since many job application packages are now submitted online, applicants do not generally have to submit written documents to the hiring agency as was common in the past

   ❑ **a.** True

   ❑ **b.** False.

3. If the hiring agency or institution requires a cover letter, it should include all of the following EXCEPT

   ❑ **a.** the vacancy number of the position you are applying for

   ❑ **b.** your contact information

   ❑ **c.** a detailed description of your educational experiences and professional goals

   ❑ **d.** a brief description of your qualifications and interest in the position advertised

4. *True or False*. The cover letter, if required, should be no longer than two pages.

   ❑ **a.** True

   ❑ **b.** False

5. All of the following should be included in your curriculum vitae EXCEPT

   ❑ **a.** your high school and university transcripts

   ❑ **b.** degrees conferred by colleges or universities with graduation dates

   ❑ **c.** publications, presentations, awards and other professional information

   ❑ **d.** employment history

6. *True or False*. If the degree you are working on has not been conferred yet, it should not be included in your curriculum.

   ❑ **a.** True

   ❑ **b.** False

7. *True or False.* It is never wise to include employment experiences that are not related to your current professional interests in a CV.

❑ **a.** True

❑ **b.** False

8. The primary purpose of a *personal statement* in a job application package is to inform the hiring institution of ___________________.

❑ **a.** personal interests such as hobbies and sports

❑ **b.** any personal or health problems that might influence your ability to perform work required by the position you are applying for

❑ **c.** computer systems used in the World Wide Web

❑ **d.** your philosophies and a discussion of any life experiences that have influenced your professional interests and career goals

9. *True or False.* The personal statement, if included in a job application, should be very brief—no more than two or three paragraphs in length.

❑ **a.** True

❑ **b.** False

10. *True or False.* Because of the efficiency of online application forms, submission deadlines for job applications are less important than they were a few years ago.

❑ **a.** True

❑ **b.** False

Answers: 1) b; 2) b; 3) c; 4) a; 5) a; 6) b; 7) b; 8) d; 9) b; 10) b.

Proposals for Grants and Contracts

Getting Your Program Funded

Chapter Learning Objectives

After studying the material in this chapter, you should understand:

- The various types of funding agencies and organizations that exist in the United States, and how they announce the availability of new funding opportunities.
- How to apply for various types of grants and contracts, and how your grant applications are processed by the funding agencies.
- How grant proposals are reviewed by agency panels, and the criteria used to rate proposals and make recommendations for or against funding.
- How to communicate effectively with program managers during the preparation of a proposal and during an active project for the purpose(s) of preparing routine progress reports, requesting budget revisions, changes in methodology, and the preparation and submission of final reports at the time the project expires.

One of the first hurdles that newly appointed faculty members must address in most research and academic institutions is obtaining funding for the research or educational programs they are expected to develop. Such funding may be available from either *internal* sources (usually the institution itself) or *external* sources such as federal or state agencies and private organizations. At any given time, funding for specific research problem areas may range from readily available to nearly nonexistent depending on the current economic conditions and political environment of the country. Nevertheless, the scientist is usually responsible for obtaining external funding, and it is therefore imperative that he or she knows how to locate funding opportunities, and how to apply for grants and contracts when suitable sources are located.

Locating Funding Opportunities

One of the most comprehensive sources of information on grants available from Federal agencies is *Grants.Gov* which lists grants and contracts that are currently available from all Federal agencies (http://www.grants.gov/web/grants/home.html). For each grant listed, a funding opportunity number and title is listed and a downloadable application package is available (Figure 11.1). Grants.Gov also provides a subscription service of funding announcements - Grants.gov Pilots (http://www.grants.gov/grants/outreach/grants-gov-pilots.html) and a Grants Learning Center (http://www.grants.gov/web/grants/learn-grants.html) that provides a series of online courses dealing with federal grants. Grants available from state governments are available on websites such as Texas Government Grants (http://www.usgrants.org/texas) and those available from private sources are available on home pages of numerous nonprofit organizations and foundations.

Governmental agencies usually announce the availability of new funding by issuing a Research Funding Announcement (RFA), Request for Proposals (RFP) or similar announcement on their websites or other public site (Figure 11.2). The announcement identifies (1) the agency involved, (2) the program title and funding opportunity number, and (3) provides additional information such as the

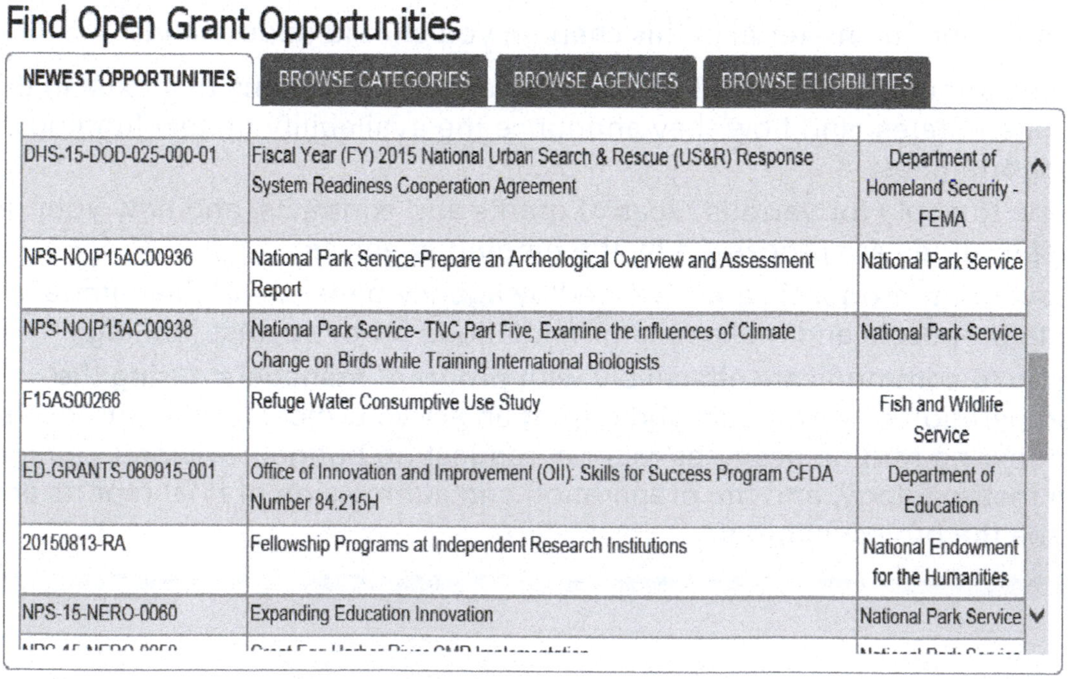

Find Open Grant Opportunities

NEWEST OPPORTUNITIES | BROWSE CATEGORIES | BROWSE AGENCIES | BROWSE ELIGIBILITIES

DHS-15-DOD-025-000-01	Fiscal Year (FY) 2015 National Urban Search & Rescue (US&R) Response System Readiness Cooperation Agreement	Department of Homeland Security - FEMA
NPS-NOIP15AC00936	National Park Service-Prepare an Archeological Overview and Assessment Report	National Park Service
NPS-NOIP15AC00938	National Park Service- TNC Part Five, Examine the influences of Climate Change on Birds while Training International Biologists	National Park Service
F15AS00266	Refuge Water Consumptive Use Study	Fish and Wildlife Service
ED-GRANTS-060915-001	Office of Innovation and Improvement (OII): Skills for Success Program CFDA Number 84.215H	Department of Education
20150813-RA	Fellowship Programs at Independent Research Institutions	National Endowment for the Humanities
NPS-15-NERO-0060	Expanding Education Innovation	National Park Service

Figure 11.1 Research Funding Announcements (RFAs) by agencies of the U.S. government are posted on Grants.Gov website.

GRANTS.GOV 〉 View Opportunity

VIEW GRANT OPPORTUNITY

« Back | Link

F15AS00266
Refuge Water Consumptive Use Study
Department of the Interior
Fish and Wildlife Service

SYNOPSIS DETAILS | VERSION HISTORY | RELATED DOCUMENTS | APPLICATION PACKAGE

Print Synopsis Details

The synopsis for this grant opportunity is detailed below, following this paragraph. This synopsis contains all of the updates to this document that have been posted as of **6/10/2015**. If updates have been made to the opportunity synopsis, update information is provided below the synopsis.

If you would like to receive notifications of changes to the grant opportunity click send me change notification emails. The only thing you need to provide for this service is your email address. No other information is requested.

Any inconsistency between the original printed document and the disk or electronic document shall be resolved by giving precedence to the printed document.

General Information

Document Type:	Grants Notice	**Posted Date:**	Jun 10, 2015
Funding Opportunity Number:	F15AS00266	**Creation Date:**	Jun 10, 2015
Funding Opportunity Title:	Refuge Water Consumptive Use Study	**Original Closing Date for Applications:**	Jun 20, 2015 This is a Notice of Intent to Make an Award to The Nature Conservancy. No competition is expected. Do not respond.
Opportunity Category:	Discretionary		
Funding Instrument Type:	Grant	**Current Closing Date for Applications:**	Jun 20, 2015 This is a Notice of Intent to Make an Award to The Nature Conservancy. No competition is expected. Do not respond.
Category of Funding Activity:	Environment		
Category Explanation:		**Archive Date:**	Jul 9, 2015
Expected Number of Awards:	1	**Estimated Total Program Funding:**	$25,000
CFDA Number(s):	15.637 -- Migratory Bird Joint Ventures	**Award Ceiling:**	$25,000
Cost Sharing or Matching Requirement:	No	**Award Floor:**	$25,000

Figure 11.2 Each funding opportunity listed in the announcement provides information regarding the grant and a downloadable application package.

amount of funding available, eligibility requirements, and proposal submission procedures and deadlines. In addition, the announcement generally includes the mission statement for the agency involved and provides the name and contact information for the program manager and other agency officials. If a potential investigator decides to submit an application for a grant, it is his or her responsibility to read the funding announcement and follow the instructions to the letter. In virtually all cases, incomplete proposals and those received by the funding agency after the stated deadline will be rejected.

Preparation of Grant Proposals

A typical proposal application contains several sections and may include a variety of attachments. The *title page* includes the title of the proposed project and contact information for the Principal Investigator (PI) and any Co-PIs that may be involved in the project (Figure 11.3). The title should be self-explanatory and brief—16 words or less is a good rule-of-thumb. The form also contains a section verifying that approval has been obtained or is pending if the proposed project includes any of the following: (1) human subjects, (2) vertebrate animals, (3) recombinant DNA, and/or (4) radioisotopes. Each of these is subject to strict regulations by federal, state, and local agencies which may require investigators to have permits and/or approval by an internal review board or similar entity. The form also contains the time frame for the proposed project, the total amount of funds requested, and is signed by the PI and officials of the requesting institution.

FACULTY RESEARCH GRANT APPLICATION
Cover page

TITLE PROJECT:
Insert–self explantory title with no more than 16 words. Capitalize each major word in title–use this same title in each section of the proposal.

Principal Investigator

New Investigator___NO___YES

Name (last, first, middle, initial):

Campus Address:

Position Title:

Department:

Telephone:

Email Address:

Check if project involves any of the following:

_____ Human Subjects
_____Vertiberate Animals
_____Recombinant DNA
_____Radioisotopes

If any of the above as checked, please indicate appropriate approval(s) have been _____ Not applied for; _____ Applied for _____ Approved on _______

Dates of Proposed Project

From October, 2009–October, 2010

COST Requested:

$ _______

Principal Investigator Assurances: I certify that the statements herein are true, complete and accurate to the best of my knowledge. I agree to accept responsibility for the compliance conduct of the project and to provide the required reports if a grant is awarded a result of this application.

Signature of PI:

Date:_______________

Signature of Chair:

Date: _______________

Signature of Dean/Director:

Date: _______________

Figure 11.3 Cover page for in-house grant application at a major university.

Virtually all proposals include an *abstract page*, which is a condensed summary of the proposed project (Figure 11.4). The abstract is typically brief (500 words or less) and should provide at least the following information: (1) the nature of the problem to be researched and why it is important to the funding agency and public, (2) objectives of the proposed research, (3) a very brief summary of the methodology proposed to accomplish the objectives, (4) the expected results and significance of the project, and (5) the duration of the project in years and the total amount of funding requested. Abstracts should always be written in future tense (since they are statements of what the investigators are proposing to do) and should not include literature citations (references will be cited in a later

RESEARCH GRANT APPLICATION
ABSTRACT

TITLE OF PROJECT: Use same title as cover page

Abstract should be no more than a single paragraph (typically 500 words or less) and should summarize the following: 1) nature of the problem to be researched and why it is important to the funding agency, 2) objectives of the proposed project (usually no more than 3 objectives), 3) a very brief discussion of methodology to be used, 4) expected results and significance of project, 4) duration of project (in years) and amount of funding requested.

Do not cite references in Abstract (references will be cited in Introduction section of Narrative section). In proposals, always write in future tense.

Figure 11.4 Abstract page for typical grant proposal in a major university.

section of the proposal). The abstract page should be written with particular care as it is usually one of the first sections of the document that are read by reviewers of the proposal.

The abstract page is usually followed by a *budget page* and *budget narrative page* (Figure 11.5). The budget page itemizes proposed costs and expenditures by categories, including salaries and wages, operating costs such as equipment, materials and supplies, consultant services, travel expenses, and other *direct costs*. The simple form used in this example does not contain calculations for employee benefits and other *indirect costs,* although these are usually components of large external grants. Each of the proposed expenditures listed in the budget page should be justified (explained) in the budget narrative page (Figure 11.6). Justifications do not have to be lengthy, but must indicate that each expense is related to the objectives of the project and conforms with the allowable expenses that were published in the original RFP. For example, a proposed $1,000 stipend to compensate undergraduate students for their services in the collection and processing of data would be justifiable in most cases, whereas a stipend to "help students complete their education" would almost never be justifiable and would probably result in rejection of the proposal by the funding agency.

The *project narrative page*(s) contains the bulk of information contained in the project proposal and consists of several distinct sections (Figure 11.7). The *introduction* section should explain (1) the nature of the problem that is to be researched and why it is important to the funding agency and general public, (2) what is known about the problem, that is, a review of pertinent literature, and (3) what is not known about it, that is, gaps in knowledge that need to be addressed. Item (3) leads logically to the *objectives* section that should normally include 2–3 tangible objectives that address significant gaps in knowledge. In the *methodology* section, the proposed procedures for each of the objectives should be explained fully and clearly, including a description of the experimental design, how experimental units will be maintained, how samples will be collected and processed, and how data will be analyzed statistically. All measurements described in the methodology section (e.g., length and weight) should be made in the metric system.

The project narrative normally includes three other sections which are unique to proposals. The section entitled *expected results* provides an opportunity for the writer to indicate to reviewers how the proposed research will address unanswered questions, and how this will benefit the funding agency and public. In this author's opinion, the best (and safest) approach here is not to speculate on results of experiments and statistical tests (i.e., ". . . we expect the mean yields of plant variety *X* to

Budget Page

PROPOSAL TITLE: Use same title as in other forms of proposal.		AWARD NO.	
		Funds Requested	Funds Granted by FRC
PERSONNEL (SHOW NUMBERS IN BRACKETS)			
(0) GRADUATE STUDENTS		$0	
(0) UNDERGRADUATE STUDENTS		$0	
(0) OTHER		$0	
TOTAL SALARIES AND WAGES		$0	
OTHER OPERATING COSTS			
1. MATERIALS AND SUPPLIES		$0	
2. PUBLICATION COSTS/ DOCUMENTATION/DISSEMINATION		$0	
3. CONSULTANT SERVICES		$0	
4. COMPUTERS SERVICES		$0	
5. OTHER		$0	
6. EQUIPMENT (LIST ITEM AND DOLLAR AMOUNT)			
equipment item 1	$0		
TOTAL EQUIPMENT		$0	
TOTAL OPERATING EXPENSES		$0	
TRAVEL 1. DOMESTIC		$0	
2. FOREIGN		$0	
TOTAL TRAVEL EXPENSES			
PARTICIPANT SUPPORT COSTS			
1. STIPENDS	$0		
2. TRAVEL	$0		
3. SUBSISTENCE	0		
4. OTHER	$0		
( ) TOTAL NUMBER OF PARTICIPANTS		$0	
TOTAL REQUEST		**$0**	

List amount of funding for each category on budget form. Salaries for undergraduate and graduate students are self-explanatory—assume that there is no indirect costs (overhead) for either. Stipends are not salaries, but are payments for other participants.

Make sure that funds for each category add up to category totals, and that category totals add up to the total amount of funding requested.

Figure 11.5 Budget page for typical grant proposal of major university.

RESEARCH GRANT APPLICATION
Budget Narrative Page

TITLE OF PROJECT:
Same title as before

Budget Explanation by Category:

For each major category, explain why the amount of funding requested is necessary.

Example:

Salaries in the amount of $2,000 are required for two undergraduate students who will be responsible for data collection, processing and analysis.

Note: don't include "equipment" and "supplies" in the same category. Supplies are expendables (paper bags used to hold samples, vials and alcohol are used to store samples of insects and mites, etc. Equipment are nonexpendable items such as microscopes, computers, etc.

Make sure that subtotals for each category add up to the total amount of funding requested.

Figure 11.6 Budget narrative provides justification statements for all major items listed in the project budget.

be significantly greater than those of variety *Y* . . . ") but to state that ". . . our research is expected to clarify differences in yields of plant varieties *X* and *Y* if such differences indeed exist. . . " Such a statement is clearly consistent with the objectives of the proposed project and does not involve speculation (and unconscious bias) on the part of the researcher(s). The section entitled *dissemination of results* describes how the researchers plan to disseminate the results of their proposed project to the funding agency, the scientific community and public in general. Most commonly, this is accomplished through periodic progress reports and final project reports to the funding agency, and oral presentations and publications in scientific journals. This section is important in the sense that virtually all funding agencies are accountable for funds awarded to researchers, and the principal measures of accomplishments are in the form of oral and written communications. A section entitled *investigator qualifications* describes the education and speciality of each of the investigators and/or collaborators listed in the proposal, and is used to determine whether or not each is qualified to carry out his or her assigned role in the proposed project. In addition to the narrative description, a recent *curriculum vitae* for each investigator is a required attachment to virtually all proposals submitted to funding agencies. The final section of most proposals is a *references* section which lists all references cited in the text of the proposal in a standard citation format.

Most funding agencies encourage grantwriters to communicate with their program managers (whose contact information is included in the funding announcement) during the preparation of proposals, and many offer the opportunity to submit *pre-proposals* prior to the submission of actual proposals. Grantwriters are wise to take advantage of these opportunities as they provide valuable feedback regarding the suitability of a given proposal as written and suggestions that might make it more competitive at the time it is submitted to a review panel.

TITLE OF PROJECT:
Same title as before

Project Narrative—This is the main body of the proposal. Here is a suggested outline—writing in outline form will make scientific writing more effective and will go a long way in avoiding "writer's block."

Introduction—Describe the nature of the problem to be addressed and its importance to the funding agency and the general public (this will be used to justify the amount of funding requested). Explain what is known about the problem (literature review) and what is unknown (the gaps in knowledge should provide the basis for your objectives). Include the important literature relating to your project and cite references.

Objectives: List the objectives of your proposed project (you should normally include 2 or 3 clearly defined objectives.
Example:
Our study will be designed to (1) identify the principal mosquito vectors of dengue occurring in the Lower Rio Grande Valley and (2) to compare the magnitude of trap catches of each species occurring in selected habitats during selected time intervals.

Methodology: Describe clearly how you will accomplish each objective.

Objective 1. Describe how you intend to collect and identify mosquito samples (cite references if necessary).

Objective 2. Describe how you will collect samples and make comparisons of trap catches in selected environments (include description of statistical tests to be used and cite references if necessary).

Expected Results: Briefly explain the results you expect to obtain from the project and its significance to the funding agency and general public. The best approach here is to indicate that your project is expected to provide significant new knowledge of an important topic (be careful about speculating about statistical results or comparisons as these are unknowns at the present time).

Dissemination of Results: Explain how you intend to distribute the results of your project to the funding agency, scientific community and general public. Most commonly this is accomplished through presentations at scientific meetings and workshops, and publication of articles in refereed journals.

Investigator Qualifications: List each investigator and briefly discuss his or her qualifications to conduct the proposed research project (you don't have to be a Bertrand Russel to get a grant, but you and the other PIs or Co-PIs need to have the training necessary to complete the project). Include a CV for each investigator.

Literature Cited: List all references cited in the Narrative section in proper format.

Figure 11.7 Project narrative contains the bulk of the research proposal.

Submission and Processing of Grant Proposals

Prior to submission, the grant application package should be proofread carefully to ensure that any typographical errors are corrected and that all forms required by the funding agency (many of which are not discussed here) are correct and are included in the package. In the recent past, most proposals were submitted to funding agencies by mailing a specified number of paper copies (commonly 10–15) to the agency official(s) listed in the funding announcement. One hazard of submitting proposals in this manner was that the submission deadlines posted by the agency generally refer to the date of receipt rather than the postmark date, and proposals arriving after the deadline were almost always rejected as required by government regulations. Today, many if not most proposals for federal and state funding are submitted online using electronic forms available on the agencies' websites. Nevertheless, the rigid submission deadlines are still enforced, and it is the scientist's responsibility to ensure that proposals submitted online are received by the funding agency on or before the deadline specified in the proposal.

Once a grant application has been received by a funding agency, it is reviewed by agency officials and by a panel of anonymous reviewers (i.e., somewhat similar to the anonymous peer review process used by most scientific journals). When evaluating proposals, reviewers typically ask themselves one or more of the following questions:

- Is the topic for the proposed research consistent with the agency's mission, and is it important enough to warrant the funding requested?
- Do the objectives address important gaps in knowledge that are critical to resolving the problem?
- Are the objectives feasible, and can they be accomplished within the time frame of the proposed project?
- If human subjects or lab animals are involved, have adequate protections been incorporated and are they in compliance with Federal and other laws?
- Are the proposed methodologies scientifically sound, and will they be evaluated by sound statistical methodologies?
- Are the investigators qualified to conduct the proposed research?
- Is the amount of funding requested sufficient to complete the objectives listed?
- How will results of the proposed project be disseminated to the public?
- Any other questions pertaining to feasibility, scientific soundness, and compliance with federal and state regulations.

Based on their evaluations, reviewers will make one of several possible recommendations: (1) reject the proposal entirely, (2) fund at a level lower than requested, or (3) fund at level requested. If the proposal is funded partially or fully, the researcher's institution will be informed and the proposed project will be activated on its scheduled start date. If the proposal is rejected, the program manager will generally forward the reviewers' comments to the researchers who may revise the proposal and submit it at a later date under most circumstances.

Dissemination of Research Results

Virtually all grants require periodic progress reports during the course of the grant and a final report when the grant project expires. Instructions for submitting periodic and final reports are outlined in the grant application package and additional information can be obtained by interacting with the program manager of the grant program (this is strongly encouraged by most funding agencies). Reports generally include a brief summary of the grant's original objectives, progress made on each objective during the reporting period, lists of presentations at scientific meetings and conferences and publications of primary and secondary literature, and any major problems encountered or significant changes made during the course of the project. Periodic and final progress reports for projects funded by federal agencies may be posted on websites accessible by other researchers and agency administrators, for example, the Current Research Information System (CRIS) maintained by the USDA National Institute of Food and Agriculture (www.cris.nifa.usda.gov). Progress reports and final reports submitted for competitive grants should be written very carefully and proofread prior to submission as they will be available to a very wide audience. For example, the CRIS system provides scientists in both the public and private sectors with information regarding the objectives of research that are being conducted by agencies or institutions funded by federal funds, the institutions and investigators involved in each project, and provides access to progress and final reports.

Scientists conducting grant-funded research also have an obligation to disseminate their research results to the scientific community and general public by means other than periodic and final progress reports to the funding agency involved. In the original project proposal, researcher(s) generally indicate their intention of disseminating their research results through a combination of oral presentations and posters at scientific meetings and conferences, and by publishing manuscripts in refereed journals and other written documents. Presentations, publications, patents, and other achievements are measures of success that funding agencies use to justify funding from state and federal legislatures, and it is therefore critical that researchers document their accomplishments and acknowledge the critical support provided by funding agencies in all oral and written communications of funded research.

Exercise 11.1

Navigate to Grants.Gov and compare current funding opportunities for different areas of the natural sciences.

Are there any notable differences in programs for chemistry, biological sciences, geology, and physics?

How does human medical research compare with environmental research?

References

Grants.gov. (2016, March 12) "*Grants Learning Center.*" http://www.grants.gov/web/grants/learn-grants.html

National Institutes of Health. (2015, June 14) "*Definitions of Criteria and Consideration for Research.*" Project Grant (RPG/R01/R03/R15/R21/R34) Critiques. http://grants.nih.gpv/grants/peer/critiques/rpg.htm

National Institutes of Health. (2015, June 14) "*Peer Review Process.*" http://grants.nih.gpv/grants/peer review process

Test Your Knowledge

Without referring to the material presented earlier in this chapter, mark an "x" in the box of the most appropriate answer to each question. Answers are provided at end of chapter.

1. Competitive grants funded by governmental agencies are announced by all of the following EXCEPT
   - ❑ **a.** RFAs
   - ❑ **b.** letters or memoranda to preferred investigators
   - ❑ **c.** RFPs
   - ❑ **d.** subscription services such as the PILOTs project

2. *True or False.* In order to promote fairness and to comply with Federal law, most competitive government grants are open to scientists affiliated with all public institutions regardless of geographic location.
   - ❑ **a.** True
   - ❑ **b.** False

3. *True or False.* During the preparation of a competitive grant proposal, the principal investigator (PI) should not contact the agency program manager as it may be interpreted as attempting to influence the outcome of the submission.
   - ❑ **a.** True
   - ❑ **b.** False

4. Certifications relating to approval to use human subjects, vertebrate animals, radioisotopes or recombinant DNA in the proposed project are included on this page of most proposals.
   - ❑ **a.** title page
   - ❑ **b.** abstract
   - ❑ **c.** budget and budget narrative
   - ❑ **d.** project narrative

5. This section of the proposal represents a condensed summary of the proposal and is usually brief (less than 500 words).
   - ❑ **a.** title page
   - ❑ **b.** abstract
   - ❑ **c.** budget and budget narrative
   - ❑ **d.** project narrative

6. *True or False*. The proposed budget should provide a "rough estimate" of various costs for the project, and minor adjustments in various categories of expenses do not generally require a formal revision of the budget.

❑ **a.** True

❑ **b.** False

7. This section contains most of the information provided in most competitive grant proposals.

❑ **a.** title page

❑ **b.** abstract

❑ **c.** budget and budget narrative

❑ **d.** project narrative

8. *True or False*. Since a section entitled *Investigator Qualifications* is included in most proposal templates, a *curriculum vitae* is not normally included in grant submission packages.

❑ **a.** True

❑ **b.** False

9. Which of the following is the primary means by which competitive grants submitted to Federal agencies are evaluated for quality, scientific soundness and potential impact?

❑ **a.** contractors

❑ **b.** agency officials

❑ **c.** selected journal editors

❑ **d.** review panels

10. *True or False*. Progress reports and final reports of most competitive government grants are filed in the highly confidential CRIS system which is available to users with appropriate security clearances.

❑ **a.** True

❑ **b.** False

Answers: 1) b; 2) b; 3) b; 4) a; 5) b; 6) b; 7) c; 8) b; 9) d; 10) b.

Letters of Recommendation and Related Documents

Services to Students, Colleagues, and Institutions

Chapter Learning Objectives

After studying the material in this chapter, you should understand:

- How to prepare effective *letters of recommendation* for students applying for admission to graduate programs and colleagues applying for jobs in the public or private sectors.
- How to prepare *employee performance evaluations* for students and other employees under your supervision at research and academic institutions.
- How to prepare *external reviews* of faculty applying for tenure and/or promotion at research and academic institutions.

Once you become an established scientist or educator with students and/or other employees working under your supervision, you will begin to receive requests from current and former students and colleagues for *letters of reference* to include in their job application packages, and you will be responsible for submitting *annual performance evaluations* for your employees. In addition, you will very probably be asked by administrators of one or more research and/or academic institutions to provide an *external review* of faculty in their institutions who have applied for tenure (permanent job status) and/or promotion. Each of these documents may affect another scientist's career, and knowing how to properly prepare each must become part of your expertise in scientific communications.

If you think that becoming an established scientist and supervising employees is a long way off and not worth worrying about right now, think again! If you are successful in your academic efforts, it will probably happen much sooner than you expect. Here is another incentive for learning this information now rather than later—the documents and procedures described in the remainder of this chapter are how YOU will be evaluated by your supervisors and peers once you land your first job. So read on!

Letters of Reference

Most job announcements posted by research and academic institutions (Chapter 10) instruct the applicant to request letters of reference from a specified number of individuals who are presumably familiar with the applicant's credentials and qualifications for the position. The applicant then has the responsibility of contacting each prospective reference and requesting a letter of reference. In general, letters of reference are submitted directly to the search committee (not to the applicant) so it is prudent for the applicant to maintain contact with both the search committee representative and the reference(s) to ensure that all letters arrive before the pending deadline.

When writing a letter of reference for a student seeking <u>admission to an academic program</u> (e.g., graduate or medical school), emphasize the following:

- **Your association with the student**—how long you have known the student and in what capacity (i.e., as instructor, as employer or supervisor, coworker),

- **An assessment of the student's academic qualifications and achievements**—prior coursework and other training in areas relevant to the program; ability to learn and understand complex concepts and facts; proficiency in both oral and written communications; any significant contributions to the scientific literature and/or to the scientific community; and any other qualifications that may affect the applicant's prospects for success in the program,

- **Personal characteristics**—motivation, dependability, and interest in chosen area of expertise; ability to work independently and as member of a team; ability to interact well with both peers and supervisors; honesty and integrity; and any other personal characteristics that may affect the applicant's prospects for success in the program,

- **Any other relevant information**—any other events or factors that could affect the applicant's performance in the program (adversely or otherwise) should be mentioned and described if appropriate,

- **A candid evaluation of the student's prospects for success in the program**—based on the information included in your letter, provide the program administrators with a candid evaluation of the applicant's prospects for success in the program. Admission of a poorly prepared or nonmotivated student to an academic program will very probably result in failure (which will be expensive and traumatic for the student and frustrating for the institution) whereas the unwarranted rejection of a well-prepared and motivated applicant will be equally frustrating and could change the career plans for a student who might have otherwise become a competent scientist.

When writing a letter of reference for a student or colleague applying for a professional position advertised by a research or academic institution, emphasize the following:

- **Your association with the applicant**—provide a summary of your interactions with the applicant, including research collaborations, grantwriting activities, and any other relevant professional activities,

- **The applicant's research and professional accomplishments** (if applicable)—provide your perspective of the applicant's major research accomplishments (including presentations, publications, patents, etc.) and their significance to the scientific community. This is particularly important for applicants applying for faculty positions emphasizing research as it is an indicator of the applicant's ability to identify problems of significance and to develop and maintain a productive research program.

- **The applicant's teaching experience and interaction with students** (if applicable)—if the applicant has previous teaching experience that you are familiar with, provide your perspective of his or her teaching expertise and accomplishments in teaching. If the applicant has no formal experience in teaching classroom courses, you might emphasize his or her ability to interact well with students in your laboratory or institution as an indicator of a potential to teach well if given the opportunity (however, if the job announcement indicates that prior teaching experience is a requirement for the position, you might advise the applicant to withdraw his or her application and apply elsewhere).

- **Personal characteristics related to success**—these characteristics are similar to those discussed previously for students—motivation, dependability, and interest in chosen area of expertise; ability to work independently and as member of a team; ability to interact well with both peers and supervisors; honesty and integrity; and any other personal characteristics that may affect the applicant's prospects for success as a faculty member in the agency or institution.

- **Any other relevant information**—any other events or factors that could affect the applicant's performance in the program (adversely or otherwise) should be mentioned and described if appropriate,

- **A candid evaluation of the applicant's prospects for success in the institution**—based on the information included in your letter, provide the search committee with a candid evaluation of the applicant's prospects for success in the institution. Hiring of a candidate who is nonmotivated or poorly prepared for the position applied for will almost certainly end in failure for the faculty member (e.g., rejection of tenure) and will be frustrating for other faculty members of the institution.

Employee Performance Evaluations

Once you are successful in obtaining internal or external funding and hire one or more employees, your workload of routine documents will increase by a factor of 1. In order to hire employees, you will be required to prepare a job announcement listing qualifications and responsibilities and post it for a specified period (see Chapters 1 and 10). After hiring, you will be responsible for evaluating the performance of each employee at routine intervals while they are working under your supervision. Performance evaluations for productive and amenable employees are usually simple and may involve checking boxes to indicate that they either met or exceeded the requirements of their positions. Evaluations for less productive and/or disruptive employees should be conducted a lot more carefully and statements explaining poor performance or disruptive activities should be written in a professional manner with documentation in order to avoid possible legal problems. Nevertheless, it is your responsibility to your other employees and to your institution to be as fair and candid in these evaluations as possible—productive employees should be rewarded and less-productive or disruptive ones should be given every opportunity to improve. If adverse personnel actions become necessary, it is imperative that you document the situation in writing in a professional manner—this is one of the few forms of writing that most scientists find distasteful.

External Reviews

It is becoming increasingly common for administrators of research and academic institutions to request *external reviews* for faculty applying for tenure and promotion. External reviews are conducted by faculty from other institutions who are considered to be experts in the applicant's discipline. In most cases, qualified external reviewers are selected from lists compiled by both the tenure/promotion applicant and the institutional administrators, and in virtually all cases, the names of external reviewers remain confidential during and after the applicant's evaluation (i.e., these are anonymous reviews). The objective of external reviews is to provide administrators with candid evaluations of an applicant's standing in the scientific community and how he or she compares with colleagues with similar job responsibilities in other institutions.

In a typical external review, the reviewer receives a dossier prepared by the applicant which includes a cover letter or statement written by the applicant, the applicant's CV, and samples of his or her scholarly material (e.g., publications) produced during the evaluation period. The reviewer then reviews and rates the candidate according to specific guidelines such as,

- the quality and impact of his or her scholarly contributions (including presentations and/or posters at scientific meetings and conferences, publications of scientific literature, patents and any other types of activities involving science) on the candidate's discipline and the scientific community in general,

- other significant contributions that the candidate has made to his discipline such as service on committees, editorial boards, task forces, and similar activities,

- in many cases, the quality of the journals he or she publishes in (journal ratings are readily available in certain online databases such as *Journal Citation Reports: Science Edition* and others).

Once the review is completed, the reviewer prepares a letter and returns it to the institution where it is placed in the candidate's folder for the remainder of the evaluation. External reviews almost never ask the reviewer to recommend approval or denial of tenure or promotion; however, they may influence the candidate's evaluation and should be prepared carefully and with due regard for the interests of both the candidate and the institution.

References

Ohio Wesleyan University. (2016, March 9) "External Review." http://www.owu.edu/about/offices-of-the-provost/faculty-personnel

Other Important Concepts

When Not to Communicate—At Least Temporarily

Classified Data, Nondisclosure Agreements, and Patent Issues

Chapter Learning Objectives

After studying this chapter, you should understand:

- Why certain types of research are considered sensitive or secret, and cannot or should not be published—at least temporarily.
- How secret or sensitive materials are disseminated to the appropriate end users.
- How to properly dissminate information involving proprietary and patentable information and data.

Research conducted during the *Manhattan Project*—the massive effort by the United States to develop the world's first nuclear weapon during World War II—was considered so secret that most of the critical activities were conducted in a few remote and highly secure locations in New Mexico, and scientists assigned to the project were prohibited from discussing any part of it with any unauthorized person(s) including spouses and other close associates. Had any critical information associated with the Manhattan Project been "leaked" or otherwise compromised, the Second World War may have had a very different ending and the world we live in now might be a very different place than it is now.

While the Manhattan Project represents an example of extreme secrecy in government-sponsored research, there are numerous types of research findings and other information that cannot or should not be released to the public—at least temporarily.

- **Classified information.** Under the authority of Executive Order 12598, the U.S. government may assign security classifications of *top secret*, *secret*, *confidential* and *restricted* to research and other information that may jeopardize the national security of the nation or cause other detrimental effects if obtained by unauthorized individuals (Bennett, 2011). The entity responsible for classifying information (*Original Classification Authorities*-OCA) must justify the level of classification and set a time limit on the duration of the classification. Under EO 12598, information cannot be classified as sensitive in order to cover up violations of the law, avoid embarrassment to various government officials or agencies or deal with any other infractions. Once documents are classified as sensitive, the OCA notifies all authorized users and access to the information is made available only to personnel with appropriate security clearances. Once the classification timeline has expired, the information is usually *declassified* and released to the public.

Example. The *CORONA* reconnaissance satellite program which was developed and operated by the U.S. government between 1960 and 1972 is an excellent example of how documents and other sensitive information are classified by the government and then declassified at a future date (Campbell & Wynne, 2011). During its 12 years of operation, *CORONA* (now known as the world's first "spy" satellite) photographed large areas of Asia with high-resolution cameras, and accumulated a huge archive of imagery that was used to assess military capabilities and industrial complexes of the Soviet Union, China and several other Asian nations. As new satellite technology developed, the *CORONA* program was eventually terminated, and its archived imagery and other information were declassified during 1995. During the height of the *CORONA* program, the high-resolution imagery and other information was classified as secret or higher and was available only to military and intelligence personnel (and their contractors) with proper security clearances. Following the declassification, *CORONA* data became available to scientists throughout the world and are now being used for a variety of purposes including change-detection analyses of urbanization, distribution of natural resources, and many other topics relating to the natural and physical sciences.

- **Government ownership.** Data and other information collected by employees of the U.S. government are the property of the U.S. government and remain so even if the employee(s) who collected them leave the government for a job in the private sector. Manuscripts and other reports prepared by government scientists and submitted for publication nearly always require review(s) by agency administrators for a variety of reasons. Former government scientists can generally obtain approval for publication of government-sponsored research (unless the information is faulty or has been classified as top secret, secret or confidential) by requesting approval to publish the data and providing a rationale to the administrators. You might also have to agree to use your own funds for page charges and to credit the agency in an acknowledgments statement. If permission is not granted, one important rule of thumb to remember is "…when you leave the government, the data stays behind."

- **Patent issues**—Individuals who discover certain types of processes or invent new and useful machines or products may be awarded a *patent*, which is a government grant of exclusive ownership of the process or product for a specified time period (U.S. Patent Office, 2014). Under U.S. Patent law (Title 35 of the United States Code), the patent holder has sole rights to manufacture and sell the product for a period of 20 years, and all others must pay royalties for its use during this period. After the patent expires, the product is considered to be *public domain* and available for anyone to use without payment of royalties.

In order to be patentable, the discovery or invention must fall into one of four categories (a process, machine, manufactured article or composition of matter and must be (1) new or novel, (2) useful, and (3) non-obvious). The "new or novel" requirement is particularly important as it clearly states that no matter how new, how novel, and how useful the product is, it will not be eligible for a patent award if it has been mentioned or described in a publication prior to the time the patent application was filed by the inventor. Remember these two important points as they save you a lot of grief:

- Never mention any information involving another scientist's research (e.g., as a *personal communication*) in a printed manuscript or document before clearing it with him or her first. An indiscretion such as this will almost certainly cost you a friend and may cost your colleague a patent award.

- If you have developed what you believe to be a patentable product, do not mention or describe it in any printed document that is available to the public before you have filed a patent application. A premature publication of your invention or discovery may help your application for tenure and/or promotion but will undermine your application for a patent.

- **Nondisclosure agreements**—Nondisclosure agreements are typically made between institutions or companies involved in research and development of certain products and individuals from outside the company who will be granted access to proprietary and other sensitive information (MOOG, 2016). The purpose of nondisclosure agreements is to utilize the expertise of outside contractors while protecting proprietary information, trade secrets, and other sensitive information for a specified period of time (e.g., 10 years). Nondisclosure agreements are legally binding and breaches may result in litigation and award(s) of damages.

Summary

Although most of the material contained in this textbook has emphasized the importance of disseminating new scientific knowledge via oral presentations and written publications, there are certain situations in which such communications may be illegal or potentially damaging to the speaker and/or writer. Present and former government scientists are generally prohibited from presenting or publishing secret or highly sensitive information in media available to the public until such information has been declassified. Nonsensitive government-owned information may be presented or published by former government scientists, but approval of agency officials is generally required even if the scientist(s) making the request collected and analyzed the original data and other information.

Publication of information relating to patentable products is done at the discretion of the scientist(s) involved, although premature publication (i.e., prior to the original filing date for a patent application) will almost certainly result in a denial of the patent application. Nondisclosure agreements are designed to protect proprietary products, trade secrets, and other information that was made accessible to an outside employee, who is prohibited from disclosing such information to other parties for a given period of time (commonly ten years). Nondisclosure agreements are legally enforceable and may involve substantial civil penalties if violated.

Scientists working with information included in any of the categories discussed previously should familiarize themselves with all of the provisions of agreements relating to their specific situations, and laws and regulations relating to proprietary information. Several of the references cited should prove helpful in this respect.

References

Bennett, J. (2011). "Classification Levels and Why Certain Information is Classified." http://news.clearancejobs.com/2011/12/25/classification-levels-and-why-certain-information (February 10, 2016)

Campbell, J. B., & Wynne, R. H. (2011). *Introduction to remote sensing* (5th ed.). New York: The Guilford Press, 667 pp.

MOOG. (2016, March 10) "Proprietary Information and Nondisclosure Agreement," http://www.MOOG.com/Literature/Corporate/Supplies/P1A-NFA.doc

U.S. Patent Office. (2014). "General Information Concerning Patents," http://www.uspto.gov/patents-getting-started /general-information-concerning-patents (February 10, 2016)

Once You Have It, Don't Lose It

Remain a Student Forever

Chapter Learning Objectives

After studying this chapter, you will understand why you should:

- Become a recognized expert, but never a guru or *prima donna*.
- Continue your productivity throughout your career.
- Maintain high professional standards for yourself and subordinates.
- Be careful about what you say and write.
- Share your knowledge—become a career builder for others.
- Remain a student always.

Most university graduates who accept appointments with academic or research institutions and maintain high professional standards and high levels of productivity during their developing careers will eventually become well-known and highly respected members of the scientific community. Once you reach one or more of your career goals, you would be well-advised to continue the practices that have contributed to your success and avoid certain pitfalls that can damage or destroy your legacy very quickly.

- **Regard yourself as an expert, but never become a guru or *prima donna*.** The ancient Greek philosopher Aristotle (384–322 B.C.) developed a view of the world (Aristotelian logic) that prevailed for nearly 2,000 years, although most of his ideas are now considered to be incomplete or incorrect. There is no such thing as an all-knowing expert in modern science, and nothing is as boring as someone who views himself or herself as a scientific guru. A real expert is one who recognizes himself as knowledgeable (perhaps the most knowledgeable) in a given subject area, but also recognizes that what he or she knows now is likely to change in the future as new scientific information becomes available.

- **Continue to be productive throughout your career**. Once they have reached a certain milestone in their careers (e.g., the award of tenure), many scientists are tempted to slow down and become somewhat lax in certain activities such as writing grant proposals, conducting novel research and publishing articles in the scientific media. Those who choose this path will usually find that their careers will plateau and become less rewarding as time passes. In contrast, those who continue to maintain productive programs will usually have plenty of competent help in the form of undergraduate and graduate students, postdoctoral associates, and other employees and colleagues for conducting research and reporting results. If you assemble a productive team, you can continue to maintain a high level of productivity with a lot less effort than was necessary earlier in your career.

- **Continue to maintain high professional standards and make sure that your students and other employees do likewise.** Although most reputable scientists do not need to be told why it is necessary to maintain high standards throughout their careers, it is equally important for them to insist that their students, postdoc associates, other employees and collaborators to do the same. The following well-publicized incident illustrates why it is imperative that senior scientists take all necessary measures to ensure that research conducted under their supervision and reported in oral presentation and scientific publications has no evidence of data fabrication, data falsification or plagiarism.

In 1986, a distinguished American virologist who was awarded a 1975 Nobel Prize coauthored a paper with a colleague who was later charged with data falsification by a government panel. The virologist was never implicated in academic misconduct himself but because of the controversy that erupted, he was forced to resign his position as President of Rockefeller University. The virologist and his colleague were eventually exonerated by a government ethics panel (1996) but only after enduring a grueling ordeal that lasted nearly five years (www.nytimes.com/books/98/09/20/specials/baltimore-scandal.html. December 4, 1991; www.encyclopedia.com February 9, 2016)

- **Be careful about what you say and write.** One of the fastest ways to destroy a reputation and career is to say or write something inflammatory in public at the wrong time. Particularly damaging are oral or written statements that are or may be interpreted as racist, sexist, or otherwise degrading to certain individuals or groups of people. The following two examples should illustrate why it is very wise to avoid inflammatory statements in any oral and written communications:

1) One of two American scientists who were awarded the 1962 Nobel Prize for identifying the double-helix structure of DNA remarked at a meeting in London that certain races of humans are less intelligent than others. Because of the uproar that ensued, the scientist was forced to resign his position as Chancellor of a major international research laboratory and retired under a cloud of controversy ("James Watson Retires After Racial Remarks," *The New York Times*, October 25, 2007, *Online*, http://www.nytimes.com/2007/10/25/science/25cnd-watson.html (2/24/2016)

2) A Professor of Economics at MIT who was instrumental in designing a major health care law in the United States stated in a videotaped panel discussion that passage of the Affordable Care Act by the U.S. Congress was due to a "lack of transparency" and to "the stupidity of the American voter." This statement resulted in a series of Congressional hearings, caused major embarrassment to a presidential administration who suddenly denied any knowledge of his existence, and severely damaged the credibility of the Professor involved (Newsmax, "Obamacare Depended on 'Stupidity of Voters,' It's Architect Says," http://nws.mx/1zeMgcm10Nov2014; www.newsmax.com/politics/obamacare (February 26, 2016).

- **Share your knowledge and become a career builder for others.** Once you become an established and respected scientist, you will have the opportunity (and obligation) to transfer your knowledge and skills to your students and junior colleagues. Remember that your success as a scientist was due in part to your ability to convey scientific knowledge to the scientific community via oral and written communications, so transferring these skills to your students should be a priority. A documented record of high-quality presentations and publications will be invaluable in helping your students gain employment with research and academic institutions, and a continuation of such productivity will be absolutely essential for remaining employed and advancing their careers with these institutions. Encourage your students to present and publish their research findings and do not feel threatened if they become as proficient as you are in speaking and writing (which they probably will). That is what you trained them to do, so take pride in it.

- **Remain a student always.** Webster's Dictionary defines a *student* as " . . . a person attending an educational institution . . . " but also as "...one devoted to careful and systematic study . . . " The first definition obviously applies to most of you reading these words right now while the second is a classical definition of a *scientist*—regardless of their formal educational status. "Remain a student always" was the theme of the author's baccalaureate ceremonies in an east Texas college some four decades ago, but is as appropriate today as it was then. Developments in science and technology are occurring at a dizzying rate, and those who "rest on their laurels" and fail to keep pace with these changes soon become "intellectual dinosaurs." Remember the old saying " . . . a growing apple is bitter and green—once it ripens, it turns red and sweet, but it is one short step away from decomposing"). On the other hand, those who remain "students" can continue to make contributions to science and the scientific community as long as they wish and are able—and that will be their legacy.

INDEX

P

R

S

W

CPSIA information can be obtained
at www.ICGtesting.com
Printed in the USA
LVOW02s1421280716
497766LV00004B/22/P

9 781465 274649